ENVIRONMENTAL RESEARCH ADVANCES

Environmental Research Advances

Peter A. Clarkson
Editor

Nova Science Publishers, Inc.
New York

LIBRARY OF CONGRESS CATALOGING-IN-PUBLICATION DATA
Environmental research advances / Peter A. Clarkson, editor.
 p. cm.
 Includes index.
 ISBN-13: 978-1-60021-762-3 (hardcover)
 ISBN-10: 1-60021-762-1 (hardcover)
 1. Environmental sciences--Research. 2. Ecology--Research. 3. Human ecology--Research. I. Clarkson, Peter A., 1957-
 GE70.E5836 2007
 577--dc22
                              2007025147

Published by Nova Science Publishers, Inc. ✦ New York

CONTENTS

PREFACE

The environment is considered the surroundings in which an organism operates, including air, water, land, natural resources, flora, fauna, humans and their interrelation. It is this environment which is both so valuable, on the one hand, and so endangered on the other. And it is people which are by and large ruining the environment both for themselves and for all other organisms. This book reviews the latest research in this field from around the globe.

Chapter 1 - Financial incentives and benefits associated with using woody biomass were the primary factors facilitating its use among the 13 users GAO reviewed. Four users received financial assistance (such as state or federal grants) to begin their use of woody biomass, three received ongoing financial support related to its use, and several reported energy cost savings over fossil fuels. Using woody biomass also was attractive to some users because it was available, affordable, and environmentally beneficial.

Several users GAO reviewed, however, cited challenges in using woody biomass, such as difficulty obtaining a sufficient supply of the material. For example, two power plants reported running at about 60 percent of capacity because they could not obtain enough material. Some users also reported that they had difficulty obtaining woody biomass from federal lands, instead relying on woody biomass from private lands or on alternatives such as sawmill residues. Some users also cited increased equipment and maintenance costs associated with using the material.

The experiences of the 13 users offer several important insights for the federal government to consider as it attempts to promote greater use of woody biomass.

Chapter 2 - The roadless areas of the National Forest System have received special management attention for decades. In part to recognize the importance of national forest roadless areas for many purposes and in part because making project decisions involving roadless areas on a forest-by-forest basis was resulting in controversy and litigation that consumed considerable time and money, the Clinton Administration established a new nationwide approach to the management of the roadless areas in the National Forest System. A record of decision (ROD) and a final rule were published on January 12, 2001, that prohibited most road construction and reconstruction in 58.5 million acres of inventoried forest roadless areas, with significant exceptions. Most timber cutting in roadless areas also was prohibited, with some exceptions, including improving habitat for threatened, endangered, proposed, or sensitive species, or reducing the risk of wildfire and disease. With some exceptions, the new prohibitions would have applied immediately to the Tongass National Forest in Alaska.

The Bush Administration initially postponed the effective date of the roadless area rule, then decided to allow it to be implemented while proposing amendments. However, the Federal District Court for Idaho preliminarily enjoined its implementation. The Ninth Circuit reversed the Idaho district court, but on July 14, 2003, the Federal District Court for Wyoming again permanently enjoined implementation of the Rule. This holding was appealed but dismissed as moot by the 10th Circuit. A final rule "temporarily" exempting the Tongass National Forest (Alaska) from the roadless rule was published December 30, 2003. On July 16, 2004, a new general roadless rule was proposed and a new final rule was published on May 13, 2005. The new rule eliminates the need for a separate Tongass rule, and replaces the Clinton roadless rule with a procedure whereby the governor of a state may petition the Secretary of Agriculture to promulgate a special rule for the management of roadless areas in that state, and make recommendations for that management. The new rule contains no standards by which the Secretary is to review a state's recommendations, and does not address the weight to be given to the views of local residents versus those of out-of-state residents with an interest in the public lands. Five states, including New Mexico and California, have filed petitions so far and those of Virginia, North Carolina, and South Carolina have been approved. Other states have until November 13, 2006, to file petitions under the special procedure or may seek a rule change thereafter under 7 C.F.R. § 1.28.

A new final forest planning rule was also published on January 5, 2005. The rule does not address roadless areas at all, but does require review of areas that might be suitable for wilderness designation when plans are revised.

This report traces the development and content of the roadless area rules and directives, describes the statutory background, reviews recent events, and analyzes some of the legal issues.

Chapter 3 - The 109th Congress, like earlier ones, continues to consider numerous policy topics that involve wetlands. Of interest are statements by the Bush Administration shortly after the 2004 election that restoration of 3 million wetland acres would be a priority. Wetland policies continue to attract congressional interest, and in recent months, that interest has become focused on the role that restored wetlands could play in protecting New Orleans, and coastal Louisiana more generally, from hurricanes.

In the 108th Congress and so far in the 109th Congress, no major wetlands legislation has been enacted. Earlier Congresses had reauthorized and amended many wetland programs and examined controversies such as applying federal regulations on private lands, wetland loss rates, implementation of farm bill provisions, and proposed changes to the federal permit program.

Congress also has been involved at the program level, responding to legal decisions and administrative actions. Examples include implementation of Corps of Engineers changes to the nationwide permit program; redefining key wetlands permit regulatory terms in revised rules issued in 2002; and a 2001 Supreme Court ruling (called the *SWANCC* case) that narrowed federal regulatory jurisdiction over certain isolated wetlands. Hearings on many of these topics were held, and some legislation was introduced. Legislation to reverse the *SWANCC* ruling has been introduced (H.R. 1356, the Clean Water Authority Restoration Act), as has a bill to narrow the government's regulatory jurisdiction (H.R. 2658, the Federal Wetlands Jurisdiction Act). A June 2006 Supreme Court ruling in two related cases could accelerate congressional attention to these issues.

Wetland protection efforts engender intense controversy over issues of science and policy. Controversial topics include the rate and pattern of loss, whether all wetlands should be protected in a single fashion, the ways in which federal laws currently protect them, and the fact that 75% of remaining U.S. wetlands are located on private lands.

One reason for these controversies is that wetlands occur in a wide variety of physical forms, and the numerous values they provide, such as wildlife habitat, also vary widely. In addition, the total wetland acreage in the lower 48 states is estimated to have declined from more than 220 million acres three centuries ago to 107.7 million acres in 2004. The long-standing national policy goal of no-net-loss has been reached, according to the Fish and Wildlife Service, as the rate of loss has been more than offset by net gains through expanded restoration efforts authorized in multiple laws. Many protection advocates view these laws as inadequate or uncoordinated. Others, who advocate the rights of property owners and development interests, characterize them as too intrusive. Numerous state and local wetland programs add to the complexity of the protection effort.

Chapter 4 - The first experiments on the use of wetland macrophytes to treat wastewaters were carried out in the Czech Republic as early as in 1970s but the first full-scale constructed wetland (CW) for wastewater treatment was built only in 1989. The last survey recognized 155 systems in operation at the end of 2002. With few exceptions all systems are designed with sub-surface horizontal flow. Most systems (91) were designed for the treatment of municipal or domestic sewage while 56 CWs were designed for the treatment of wastewater from combined sewer systems. Size distribution of constructed wetlands varies between 3 and 1200 PE (population equivalent) with most systems being designed for the size between 100 and 500 PE (60 systems) followed by small on-site systems for less than 10 PE (43 systems). As most CWs are designed with the specific area of about 5 m^2 per one population equivalent the most frequent size of vegetated beds is between 500-2500 m^2 (67 systems) and < 50 m^2 (42 systems). The area of vegetated beds varies between 9 and 5630 m^2. The most frequently used macrophyte is Common reed (*Phragmites australis*) which is used in 42 systems as a monotypic stand and in 50 systems in combination with other macrophytes such as Reed canarygrass *(Phalaris arundinacea)* or Cattails (*Typha* spp.). Constructed wetlands are very effective in removing organics BOD_5, COD, suspended solids and bacteria. Removal of BOD_5 and suspended solids is usually > 90% and removal of bacteria is commonly > 99%. Removal of nutrients is lower not exceeding 50% for both nitrogen and phosphorus. Constructed wetlands in the Czech Republic proved to be a suitable alternative to conventional treatment systems for small sources of pollution.

Chapter 5 - The 109th Congress is considering legislation and conducting oversight on National Park Service (NPS) related topics. The Administration is addressing park issues through budgetary, regulatory, and other actions. Earlier Congresses and Administrations also have dealt with similar issues. While this report focuses on several key topics, others may be added if circumstances warrant.

Historic Preservation. The NPS administers the Historic Preservation Fund (HPF), which provides grants to states and other entities to protect cultural resources. Congress provides annual appropriations for the HPF, and views differ as to whether to retain the federal role in financing the fund. Legislation to reauthorize the HPF (S. 1378 and H.R. 5861) is being considered. Further, the Advisory Council on Historic Preservation has issued a draft revision of its policy statement regarding treatment of burial sites, and the draft has been controversial.

Maintenance Backlog. Attention has focused on the NPS's maintenance backlog, estimated by DOI at between $5.80 billion and $12.42 billion for FY2005. Views differ as to whether the backlog has increased or decreased in recent years, and the NPS has been defining and quantifying its maintenance needs. H.R. 1124 and S. 886 seek to eliminate the NPS maintenance backlog and the annual operating deficit.

Policy Revisions. The NPS has revised its service-wide management policies — one of the authorities governing decision-making on a wide range of issues. The final policies, issued August 31, 2006, dropped many of the proposed changes that were controversial. The House and Senate have held hearings on this issue, related NPS authorities, and broader management issues.

Wild and Scenic Rivers. The Wild and Scenic Rivers System preserves free-flowing rivers, which are designated by Congress or through state nomination with Secretarial approval. The NPS, and other federal agencies with responsibility for managing designated rivers, prepare management plans to protect river values. Management of lands within river corridors is sometimes controversial, because of a variety of issues including the possible effects of designation on private lands and of corridor activities on the rivers. Legislation is pending to designate, study, or extend components of the system, and some of these measures have passed the Senate or House.

Other Issues. Some other park management topics of interest to the 109th Congress are covered here. They relate to the competitive sourcing initiative, whereby certain NPS activities judged to be commercial in nature are subject to public-private competition; air quality at national park units; and security of park units, particularly at national icons and along international borders.

Chapter 6 - Although the authors have not yet completed their review, their preliminary observations are that EPA did not adhere to all aspects of its rulemaking guidelines when developing the new TRI reporting requirements. EPA's Action Development Process outlines a series of steps to help guide the development of new environmental regulations. Throughout this process, however, the senior EPA management has the authority to accelerate the rule development process. Nevertheless, while the authors continue to pursue a clearer understanding of EPA's actions, they have identified several significant differences between the guidelines and the process EPA followed in this case: (1) late in the rulemaking process, senior EPA management directed consideration of a burden reduction option that the TRI workgroup had previously dropped from consideration; (2) EPA developed this option on an expedited schedule that appears to have provided a limited amount of time for conducting various impact analyses; and (3) EPA's decision to expedite Final Agency Review, when EPA's internal and regional offices determine whether they concur with the final proposal appears to have limited the amount of input they could provide to senior EPA management. First, the TRI workgroup charged with identifying options to reduce reporting burdens on industry identified three possible options for senior management to consider. The first two options allowed facilities to use Form A in lieu of Form R for PBT chemicals, provided the facility has no releases to the environment, and the third created a "no significant change" reporting option in lieu of Form R for facilities with releases that changed little from the previous year. Information from a June 2005 briefing for the Administrator indicated that, while the Office of Management and Budget (OMB) had suggested increasing the Form A eligibility for non-PBT chemicals from 500 to 5,000 pounds, the TRI workgroup dropped that option from consideration.

Second, although the authors could not determine from the documents provided by EPA what actions the agency took between the June 2005 briefing for the EPA Administrator and the October 2005 issuance of the TRI proposal in the *Federal Register*, the Administrator provided direction after the briefing to expedite the process in order to meet a commitment to OMB to provide burden reduction by the end of December 2006. Subsequently, EPA revised its economic analysis to include consideration of the impact of raising the Form A eligibility threshold. However, that analysis was not completed before EPA sent the proposed rule to OMB for review and was only completed just prior to the proposal being signed by the Administrator and published in the *Federal Register* for public comment.

Third, the extent to which senior EPA management sought or received input from internal stakeholders, including the TRI workgroup, after directing reconsideration of the option to increase the Form A reporting threshold from 500 to 5,000 pounds for non-PBT chemicals remains unclear. The authors have been unable to determine the extent to which EPA's internal and regional offices had the opportunity during Final Agency Review to determine whether they concurred with the proposal to increase the Form A threshold. They will continue to pursue the answer to this and other questions as we complete our work. Finally, in response to the public comments on the proposal, nearly all of which were negative, EPA considered alternative options and revised the proposal, thereby allowing facilities to report releases of up to 2,000 rather than 5,000 pounds on Form A.

The authors believe that the TRI reporting changes will likely have a significant impact on information available to the public about dozens of toxic chemicals from thousands of facilities in states and communities across the country. EPA estimated that the TRI reporting changes will affect reporting on less than 1 percent of the total chemical releases reported to the TRI annually. While their analysis supports EPA's estimate of this impact at a national level, it also suggests that changes to TRI reporting requirements will have a significant impact on the amount and nature of toxic release data available to some communities, information that is ultimately much more meaningful to citizens. In addition, preliminary results from our January 2007 survey of state TRI coordinators indicates that as many as 23 states believe that EPA's changes to TRI reporting requirements will have a negative impact on various aspects of TRI. To develop a more specific picture of the impact of the TRI reporting changes at a local level, we used 2005 TRI data to estimate, by state, the impact of EPA's changes. First, they estimated that the detailed information from more than 22,000 Form R reports may no longer be included in the TRI if all eligible facilities begin using Form A. More specifically, Alaska, California, Connecticut, Hawaii, Massachusetts, New Jersey, and Rhode Island could have 33 percent fewer chemical reports. Second, they estimated that the number of chemicals for which no information could be reported under the new rule ranges from 3 chemicals in South Dakota to 60 chemicals in Georgia. Thirteen states—including Delaware, Georgia, Maryland, Missouri, Oklahoma, Tennessee, and Vermont—could have no detailed reports on more than 20 percent of reported chemicals. Third, they estimated that a total of 3,565 facilities would no longer have to report quantitative information about their chemical use to the TRI. In fact, more than 20 percent of facilities in Colorado, Connecticut, Hawaii, Massachusetts, and Rhode Island, could have no detailed information about their chemical use. Furthermore, citizens living in 75 counties in the United States—including 11 in Texas, 10 in Virginia, and 6 in Georgia—could have no quantitative TRI information about local toxic pollution. Finally, with regard to the impact of the rule change on industry's reporting burden, EPA estimated that, if all eligible facilities take advantage of the reporting changes, they could save a total of about $5.9 million—about 4 percent of the annual cost of TRI reporting. This is the equivalent of less than $900 per facility. However, based on past experience, not all eligible facilities will use Form A, so the actual savings to industry are likely to be less.

It should be noted that perchlorate releases are not reported to the TRI. Perchlorate, a primary ingredient in propellant, has been used for decades in the manufacture and firing of rockets and missiles. Other uses include fireworks, flares, and explosives. Perchlorate is a salt that is easily dissolved and transported in water and has been found in groundwater, surface water, drinking water, soil, and food products such as milk and lettuce across the country. Health studies have shown that perchlorate can affect the thyroid gland and may cause developmental delays. The authors identified more than 400 sites in 35 states where perchlorate had been found in concentrations ranging from 4 parts per billion to more that 3.7 million parts per billion, and that more than one-half of the sites were in California and Texas. However, federal and state agencies are not required to routinely report perchlorate findings to EPA, and EPA does not centrally track or monitor perchlorate detections or the status of cleanup efforts. As a result, a greater number of contaminated sites than the authors reported may exist. Although concern over potential health risks from perchlorate has increased, and at least 9 states have established nonregulatory action levels or advisories, EPA has not established a national drinking water standard citing the need for more research on health effects. They concluded in our report that EPA needed more reliable information on the extent of sites contaminated with perchlorate and the status of cleanup efforts, and recommended that EPA work with the Department of Defense and the states to establish a formal structure for tracking perchlorate information. In December 2006, EPA reiterated its disagreement with the recommendation stating that perchlorate information already exists from a variety of other sources. However, they continue to believe that the inconsistency and omissions in available data that we found during the course of our study underscore the need for a more structured and formal tracking system.

Chapter 7 - High density plantations of fast-growing tree species grown on fertile land are today a viable alternative for the production of bio-fuels in many countries of cool-temperate regions. For example in Sweden, willow biomass plantations are commercially grown for energy purpose. Rapid development of sustainable production systems and infrastructure, along with recent progress in tree crop breeding resulted in high biomass production potential on fertile agricultural land. In addition, tree plantations can increase biodiversity in open agricultural landscapes and serve as tools for the amelioration of environmental problems at local (e.g., waste product contamination through phytoremediation) and global scale (e.g., increased greenhouse effect through carbon sequestration). Thus, multifunctional biomass plantations offer additional possibilities in terms of, e.g., wastewater cleaning and carbon sequestration. However, the establishment and large-scale implementation of woody biomass plantations is often controversial due to, for example, presumed negative influences on biodiversity and the cultural heritage landscape, along with negative public attitudes. Conflicting interests of different parts of society and socio-political issues (e.g., agricultural and energy policy, market developments, public attitudes) are therefore major barriers for the rapid development of woody bio-fuel plantations in Sweden and many other countries, rather than climatic, technical or environmental constraints. In future, careful analysis of the non-technical, non-climatic barriers at regional level and possibly the development of guidelines for the establishment and sustainable, environmentally friendly management of woody biomass plantations could be means to boost the utilization of woody biomass from agricultural land in many countries of the world.

Chapter 8 - The modern life has increased the sources from which particles and gases could be liberated into the atmosphere. The increase in fuel combustion as a consequence of the augmentation in vehicular transportation, wood burning, land erosion, wheel and asphalt erosion, and industrial burning products are some of the origins of these aerosols and gases. Also, smoking should be mentioned as an important source of atmospheric pollution.

Atmospheric aerosols are complex mixtures of particles directly emitted into the atmosphere and particles that are formed during gas-to-particles conversion process (Kourtrakis and Sioutas, 1996).

Chapter 9 - Critical limits for cadmium in the human food chain are considered by some to have too small margins of safety and some are regularly exceeded. There is therefore widespread concern about the exposure of some sections of the population to cadmium in the human food chain, in particular from offal, which is a major depot for cadmium in livestock products. During the last decade a large number of empirical models and only a very few mechanistic-phenomenological models for food chain Cd accumulation were developed.

Linking sector-specific models for Cd accumulation in the food chain is a challenge as a large variety of modelling techniques are usually used to derive specific sub-models. The easiest models to integrate into large food-chain models are the linear and nonlinear empirical models. The authors used several modeling techniques for sub-model construction, but in particular meta-regression for compiling results from many trials conducted on the absorption of Cd from soil by plants, on the uptake of Cd into sheep kidney and liver, and on the relationship between Cd intake by humans and standard measures of human toxicity.

Human cadmium intake derives mainly from food sources, although cigarette smoking is a significant source. Cadmium can be present in high concentrations in some offal and accumulates in kidney and liver tissue, with a greater rate for kidney accumulation than for liver. The margin of safety between typical cadmium intakes by humans and levels associated with toxicity is smaller than for other heavy metals. The authors determined that there is an exponential increase in β2-microglobulin with increases in cadmium intake above 302 μg/day, which corresponds to a Provisional Tolerable Weekly Intake of 3.02 μg/kg of body weight, when a Safety Margin of 10 is included. This compares with the current level set by FAO/WHO of 7 μg/kg of body weight, which is therefore believed to be too high. Human cadmium intake could be effectively reduced by routine removal of the ovine liver and kidney from the food chain in polluted regions. Consumption of just one sheep kidney could cause an average adult person to exceed their Provisional Maximum Tolerable Daily Intake (1μg/kg body weight).

The meta-analysis of published data results was used to derive prediction equations of cadmium accumulation in sheep which are applicable to a broad set of exposure situations allowing the critical examination of cadmium in the human food chain. The product of the cadmium concentration in the feed and the duration of exposure to that feed were significant predictors of the cadmium concentration in liver and kidney. The predominance of organic rather than inorganic forms of cadmium in the feed further increased accumulation. As a result, the prime measure to decrease the risk of cadmium from animal origin adversely affecting human health should be restricting the cumulative cadmium intake of herbivores. Since reduction of maximum cadmium levels in sheep feed or of the duration of their exposure are not economically viable measures of control, routine removal of the ovine liver and kidney from the food chain in polluted regions is recommended as the best option to reduce human cadmium intake. In abattoirs, cadmium concentrations in farm animals are best monitored from hair and blood concentrations, respectively.

The meta-analyses of soil data showed that a particular care should be taken if a soil has both a high soil cadmium concentration and a low pH because this situation will favour cadmium accumulation in plants. The use of soil models for monitoring cadmium fate revealed that Cd^{2+} and Pb^{2+} concentrations in the soil solution phase of contaminated soils may be reasonably well predicted from the total soil-metal concentration and the soil pH. The amounts of cadmium taken up by plants or lost through leaching are relatively small compared with the amount added to soils through phosphate fertilizer application. There is therefore a net accumulation of cadmium in soils, particularly in the upper 15 cm. The maximum soil Cd concentrations quoted by international guidelines vary between 1 mg/kg and 3 mg/kg and pasture soils contain on average 0.4 mg Cd. Some fertilizers companies have adopted the limit of 280 mg Cd/kg P, representing 25 mg Cd/kg fertilizer in the case of single superphosphate.

Linear modelling of field/pot plant data enabled us to draw inferences on Cd accumulation in plants. There is a linear relationship between the total Cd concentration in the vegetative plant (dependent variable) and the plant dry weight, the total Cd concentration in the soil solution and the duration of exposure (time from planting to harvest) (independent variables), as well as a linear relationship between Cd concentration in the reproductive organs /grains and Cd concentration in the vegetative plant organs. No linear relationship was found between Cd concentration in bulk soil and plant organs, or between Cd concentration in the soil solution and grains/reproductive organs. Although Cd uptake influences the total concentration of Ca, Mg, K, P, Mn, Fe, Cu in plants, especially in the early stages of their development, the total Cd concentration in the vegetative plant part is not influenced by the concentration of these metals in the respective organs. For non-toxic Cd concentration levels in the plant organs, the plant mineral status does not influence Cd uptake.

Mechanistic models have been constructed to estimate mineral root uptake, which usually assume that ions are transported to roots by mass flow and diffusion and are absorbed at rates that depend on their concentration at the root surface, following Michaelis-Menten kinetics. The accuracy of such models is not high, partly because accurate estimates of root surface area are essential but difficult to achieve. Other important parameters are the cadmium concentration in the soil solution and root growth.

Soil temperature variation is considerable but little attention has been paid to this factor, which has a major impact on cadmium influx into the plant. Other factors, such as the presence of competing ions, sewage sludge use as a fertilizer, symbiotic fungi (arbuscular mycorrhizae), and bacteria, and temperature are believed to play an important role in cadmium uptake but are mainly unquantified. Soil splash may play an important role in the contamination of crops which have the edible parts within 30-40 cm above the soil surface. The contribution of wet and dry deposition also plays an important role in heavy metal accumulation in plants, but little attention has been paid to cadmium accumulation throughout the phylloplane.

There is currently only a limited understanding and quantification of key parameters which would allow a comprehensive mechanistic model of Cd uptake by different plant genotypes to be constructed, and also that there is a limited number of empirical observations of key endpoints for an empirical model. Further work is required to understand each species' influence on plant model parameters. Survey data of field-grown plants might help to elucidate different aspects of the relationship between soil parameters, air depositions and cadmium accumulation in plants.

In order to predict cadmium and lead concentrations in air, there are well-established published models that can be used. These models take into account the complex phenomena in the atmosphere and they can make estimations of the deposition rates for Cd and Pd from data on source type, the intensity of atmospheric turbulence, the emission rate, the wind speed, the degree of phenomenon complexity, the surface structure and vegetation properties of the reception area.

Further work on these aspects is essential to facilitate the construction of effective models to control excessive Cd accumulation in the human food chain.

Chapter 10 - An adequate level of dissolved oxygen is necessary to support most forms of aquatic life. Very low levels of dissolved oxygen (hypoxia) in bottom-water *dead zones* are natural phenomena, but can be intensified by certain human activities. Hypoxic areas are more widespread during the summer, when algal blooms stimulated by spring runoff decompose to diminish oxygen. Such hypoxic areas may drive out or kill animal life, and usually dissipate by winter.

The largest hypoxic area affecting the United States is in the northern Gulf of Mexico near the mouth of the Mississippi River, but there are others as well. Most U.S. coastal estuaries and many developed nearshore areas suffer from varying degrees of hypoxia, causing various environmental damages. Research has been conducted to better identify the human activities that affect the intensity and duration of, as well as the area affected by, hypoxic events, and to begin formulating control strategies.

Near the end of the 105th Congress, the Harmful Algal Bloom and Hypoxia Research and Control Act of 1998 was signed into law as Title VI of P.L. 105-383. Provisions of this act authorize appropriations through NOAA for research, monitoring, education, and management activities to prevent, reduce, and control hypoxia. Under this legislation, an integrated Gulf of Mexico hypoxia assessment was completed in the late 1990s. In 2004, Title I of P.L. 108-456, the Harmful Algal Bloom and Hypoxia Amendments Act of 2004, expanded this authority and reauthorized appropriations through FY2008.

As knowledge and understanding have increased concerning the possible impacts of hypoxia, congressional interest in monitoring and addressing the problem has grown. The issue of hypoxia is seen as a search for (1) increased scientific knowledge and understanding of the phenomenon, as well as (2) cost-effective actions that might diminish the size of hypoxic areas by changing practices that promote their growth and development. This report presents an overview of the causes of hypoxia, the U.S. areas of most concern, federal legislation, and relevant federal research programs.

Chapter 11 - The National Park System includes 390 diverse units administered by the National Park Service (NPS) of the Department of the Interior. Units generally are added to the National Park System by act of Congress, although the President may proclaim national monuments on land that is federally managed for inclusion in the system. Before enacting a law to add a unit, Congress might first enact a law requiring the NPS to study an area, typically to assess its national significance, suitability and feasibility, and other management options. Important areas also are preserved outside the National Park System through programs managed or supported by the NPS.

Chapter 12 - The Environmental Quality Incentives Program (EQIP) provides farmers with financial and technical assistance to plan and implement soil and water conservation practices. EQIP was enacted in 1996 and amended by the Farm Security and Rural Investment Act of 2002 (Section 2301 of P.L. 107-171). It is a mandatory spending program (i.e., not subject to annual appropriations) administered by the Natural Resources Conservation Service. EQIP is guaranteed a total of $10.0 billion from FY2002 through FY2010 from the Commodity Credit Corporation (CCC), making it the largest conservation financial assistance program [1]. Issues about EQIP that Congress may explore as it starts to consider the next farm bill include (1) reducing the pending backlog of applications, (2) measuring the program's accomplishments, and (3) using EQIP to address specific topics or needs in specified locations.

In: Environmental Research Advances
Editor: Peter A. Clarkson, pp. 1-3

ISBN: 978-1-60021-762-3
© 2007 Nova Science Publishers, Inc.

Expert Commentary A

HUMAN HEALTH: THE GREAT FORGOTTEN AFTER ACCIDENTAL OIL SPILLS

Blanca Laffon[1,2,], Beatriz Pérez-Cadahía[1,2], Eduardo Pásaro[1] and Josefina Méndez[2]*

[1]Toxicology Unit, University of A Coruña, Edificio de Servicios Centrales de Investigación, Campus Elviña s/n, 15071-A Coruña, Spain
[2]Dept. Cell and Molecular Biology, University of A Coruña, Faculty of Sciences, Campus A Zapateira s/n, 15071-A Coruña, Spain

Since industrial revolution took place in the 18th century, use of fossil fuels, especially petroleum derivatives, has been increasing more and more. This requires their transport from the platforms where they are extracted to everywhere around the world, usually performed through sea routes in big tankers. The bad state of a considerable number of them, added to the fact that many are still monohull, is closely related to the high number of accidental spills that occurred in the last decades.

The main ecosystem components affected by the spills are generally seaside flora and some fauna species as birds and bivalve mollusks. Nevertheless, when a big spill occurs there is usually a numerous group of voluntary people, composed generally by zone inhabitants, who mobilizes and takes part in the cleaning labors, in order to minimize the impact of the spill on the natural an economic resources and recover the coastal environment as soon as possible. These individuals constitute an exposed population whose health may be potentially affected by the noxious properties of the oil.

Harmful effects of oil spills on diverse marine species, especially birds and marine invertebrates, have been extensively studied. It is enough to introduce the name of any sunken oil tanker (e.g. Exxon Valdez, Nakhodka, Erika, Prestige, etc.) in a bibliographic search engine (e.g. PubMed) and many studies on the impact of the spill on coastal ecosystems and the contamination and recovery characteristics are obtained.

Nevertheless, there are only a few studies on the repercussions of oil exposure on human health. Most of them are related to acute effects such as headache, low back pain, leg pain,

---

* Corresponding author. Tel.: +34 981 167000; fax: +34 981 167172. E-mail address: blaffon@udc.es

and eyes and respiratory irritations [1-4], and to psychological symptoms as depression, anxiety or posttraumatic stress [2, 5, 6]. So far there has been a practical absence of studies dealing with chronic health effects. This acquires a particular relevance taking into account that the main groups of substances in which the evaluation of the potential risk caused by the oil must be focused (volatile organic compounds, polycyclic aromatic hydrocarbons and heavy metals) [7-9] are mainly characterized by their carcinogenic properties. Therefore, it seems reasonable to suggest genotoxicity assays as those that best fit for the purpose of evaluating human chronic toxicity of oil exposure. But, what is the reason that underlies this absence? Perhaps the scientific community has lacked in interest on this topic until now, but one can also wonder if decision-making bodies involved in one way or another in the government of those countries having large extensions of coastal territories, should not impel the development of this kind of studies, both by encouraging specialized research groups and by providing with financial support.

On the other hand, considering the relatively high frequency of this kind of environmental disasters, it seems clearly necessary to have at general disposal detailed intervention protocols established following technical criteria and independent of political opinions. These protocols should estipulate the most adequate action for each case (e.g. to move a tanker having a mishap close to a port or to move it further away from the coast instead). Moreover, they should include some mechanisms to detect and control the possible harmful health effects that exposure can induce, just as to perform the immediate collection of biological samples from the beginning of the cleaning works, in order to establish as the levels of internal individual exposure as the effects at the chronic level, especially those related to genotoxicity. This will permit not only to determine the risk that exposure may involve, but also to evaluate if protective devices used by the subjects in each case fulfilled adequately their function, or on the contrary they did not exert the required protection and therefore it is necessary to revise material characteristics and their instructions for use.

REFERENCES

[1]   Attias L, Bucci AR, Maranghi F, Holt S, Marcello I, Zapponi GA. Crude oil spill in sea water: an assessment of the risk for bathers correlated to benzo(a)pyrene exposure. *Cent. Eur. J. Public Health.* 1995;3;:142-5.

[2]   Lyons RA, Temple JMF, Evans D, Fone DL, Palmer SR. Acute health effects of the Sea Empress oil spill. *J. Epidemiol. Community Health.* 1999;53:306-10.

[3]   Morita A, Kusaka Y, Deguchi Y, Moriuchi A, Nakanaga Y, Iki M, et al. Acute health problems among the people engaged in the cleanup of the Nakhodka oil spill. *Environ. Res.* 1999;81:185-94.

[4]   Suárez B, Lope V, Pérez-Gómez B, Aragonés N, Rodríguez-Artalejo F, Marqués F, et al. Acute health problems among subjects involved in the cleanup operation following the Prestige oil spill in Asturias and Cantabria (Spain). *Environ. Res.* 2005;99:413-24.

[5]   Palinkas LA, Petterson JS, Russell J, Downs MA. Community patterns of psychiatric disorders after the Exxon Valdez oil spill. *Am. J. Psychiatry.* 1993;150:1517-23.

[6]   Palinkas LA, Russell J, Downs MA, Petterson JS. Ethnic differences in stress, coping, and depressive symptoms after Exxon Valdez oil spill. *J. Nerv. Ment. Dis.* 1992;180 :287-95.

[7]   Commission Directive 93/67/EEC of 20 July 1993 laying down the principles for assessment of risks to man and the environment of substances notified in accordance with Council Directive 67/548/EEC. *Official Journal L* 084, 0009-0018; 08/09/1993.

[8]   Commission Regulation (EC) No. 1488/94 of 28 June 1994 laying down the principles for assessment of risks to man and the environment of existing substances in accordance with Council Regulation (EEC) No. 793/93 (Text with EEA relevance). *Official Journal L* 161, 0003-0011; 29/06/1994.

[9]   Council Regulation (EEC) No. 793/93 of 23 March 1993 on the evaluation and control of the risks of existing substances. *Official Journal L* 084, 0001-0075; 05/04/1993.

In: Environmental Research Advances
Editor: Peter A. Clarkson, pp. 5-17

ISBN: 978-1-60021-762-3
© 2007 Nova Science Publishers, Inc.

Expert Commentary B

ALTERNATIVE STATES AND TEMPORARY WETLANDS: RESEARCH OPPORTUNITIES FOR UNDERSTANDING EFFECTS OF ANTHROPOGENIC STRESS AND NATURAL DISTURBANCE

David G. Angeler[a,], Andrew J. Boulton[b], Kim M. Jenkins[b],
Beatriz Sánchez[a], Miguel Alvarez-Cobelas[c] and
Salvador Sánchez-Carrillo[c, d]*

[a]University of Castilla – La Mancha,
Institute of Environmental Sciences (ICAM), Avda. Carlos III s/n,
E-45071 Toledo, Spain
[b]University of New England, Ecosystem Management,
Armidale, 2351 NSW, Australia
[c]CSIC, Instituto de Recursos Naturales, Serrano 115 dpdo,
E-28007 Madrid, Spain
[d]Instituto Tecnológico de Sonora, Dept. of Water Sciences and Environment,
5 de febrero 818 sur, Ciudad Obregón, Sonora, 85000 México

INTRODUCTION

The introduction of alternative stable state (ASS) theory to ecology in the 1970s (Lewontin 1969) had an immense influence on basic and applied ecological research in terrestrial and aquatic ecosystems (Beisner et al. 2003; Suding et al. 2004; Didham et al. 2005; Schröder et al. 2005). The ASS concept posits that ecological systems may shift between contrasting states, the shifts being triggered either by altered community structure in similar abiotic environments (Chase 2003a, b) or by changing abiotic environmental settings that cause major community shifts (e.g., Scheffer et al. 2001; Dent et al. 2002). What are the ecological and management implications of these shifts across contrasting states? Do they

* Corresponding author: david.angeler@uclm.es

occur consistently? Are the states relatively stable? What are the main factors that drive these changes and how do the biotic and abiotic drivers interact?

In aquatic ecology, the ASS paradigm is based on research results from north-temperate, shallow lakes where cultural eutrophication is the main driver causing lakes to shift from the clear-water, submerged macrophyte-dominated state to the degraded, turbid, phytoplankton-dominated state (e.g., Scheffer et al. 1993; Jeppesen et al. 1997, Figure 1). This classic example from shallow lakes is an ideal starting point from which to reflect on the current status of the ASS concept in aquatic ecology. Our most pressing environmental problems demand consideration of diverse anthropogenic impacts that act in concert with global climate change, potentially pushing ecosystems irreversibly to other alternative states (Falk et al. 2006). This raises major challenges in managing novel ecosystems in ways that go beyond the simple control of algal blooms and includes maintaining sustainability that guarantees the provision of ecosystem services to humans (Kremen 2005).

To address and understand the generalities of ecological patterns associated with shifts in state and context-dependent feedback effects, it is necessary to include a wider range of aquatic ecosystem types. We need to extend beyond the classic model based on shallow lakes and consider other abundant wetland types, particularly temporary waters, one of the most widespread wetland types on the globe (Comín and Williams 1994).

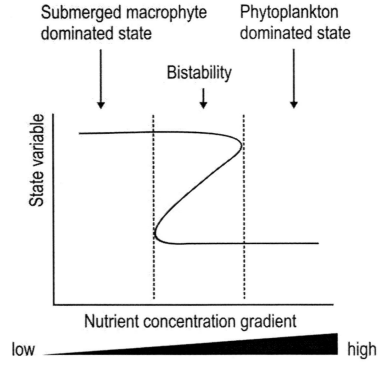

Figure 1. Model showing the ASS concept using shallow lakes as an example and eutrophication (environmental variable) as the driving force that induces state shifts. At low nutrient concentrations, lakes exist in the macrophyte-dominated state while at high nutrient concentrations, lakes occur in the phytoplankton-dominated state. At intermediate nutrient concentrations, both states can occur alternatively (bistability).

Evidence is mounting that temporary waters can exist in ASSs (Cottenie et al. 2001; Kingsford et al. 2004; Ruggiero et al. 2005; Angeler and Moreno 2006; DeClerk et al. 2006; Jenkins and Boulton in press) so our aim here is to review several case studies of temporary waters to highlight the complex ecological behaviour of these ecosystem types, their ASSs, and the implications of these aspects to successful management and wetland restoration in a swiftly-changing world.

WHY TEMPORARY WETLANDS?

Not only are temporary wetlands widespread globally, they typify semi-arid and arid areas where water is scarce and so wetland resource problems are most severe. Temporary waters span an extraordinary range of different ecosystem types, including saline and freshwater lakes, pools and ponds that lie along a water permanence gradient from ephemeral to permanent, and encompass arid and semi-arid rivers with their associated floodplains (Williams 2006). Although highly variable in shape, permanence, geological setting, water quality and community composition (often with unique and rare flora and fauna), all temporary wetlands share a common feature that distinguishes them clearly from north-temperate shallow lakes: the strong abiotic (natural disturbance) control of ecosystem organization that leads to marked intra- and inter-annual fluctuations of their communities and ecosystem processes (Mitsch and Gosselink 2000). Therefore, they are ideal 'field laboratories' for studying the interactive effects of natural disturbance by drying and anthropogenic stress, a topic that has been neglected in applied limnology, but with important repercussions for effective ecosystem management and restoration.

Many temporary wetlands are biodiversity 'hotspots', providing fundamental ecosystem services to wildlife (e.g., breeding and foraging habitat) and humans (flood protection, aesthetic and recreational values) (Williams 2006). Yet their ecology is poorly-studied and many examples of these little-known ecosystems are disappearing at an increasing rate due to agricultural expansion, alteration of flow regimes, draining and damming, and habitat fragmentation (Robinson 1995; Gibbs 2000, Kingsford 2000a). These threats are likely to intensify with climate change so their sustainable use and effective management is therefore all the more critical (Jenkins et al. 2005). By exploring some case studies of some of the few well-studied temporary wetlands, we seek common features associated with ASSs that may help guide such effective management.

CASE STUDIES

1. Artificial Temporary Ponds and Fire Retardant Contamination

Fire retardants used for wild-fire prevention and extinguishing operations can cause eutrophication of surface waters. Angeler and Moreno (2006) measured water quality in constructed outdoor ponds over three hydroperiods to determine impact-recovery trajectories in retardant-contaminated, temporary wetlands. They used a multiple before–after control–impact (MBACI) design to determine the effects of application rates that are used in grasslands (1 L m^{-2}) and scrublands (3 L m^{-2}). Retardant application caused a significant increase in the trophic status of the ponds in the post-contamination period (second and third hydroperiod) relative to the pre-contamination period (first hydroperiod), resulting in a shift from clear water to turbid water stable states that persisted for the remainder of the study.

In a subsequent study, Angeler and Moreno (in press) studied zooplankton community recovery in retardant ponds. The retardant caused a decline in species richness and an increase in rotifers during summer and winter months relative to controls and pre-treatment dates, and the duration of these changes varied among retardant treatments. When trajectories of change in rotifer densities were plotted on nonmetric, multidimensional scaling ordinations, they followed loops that showed recurring deviations from, and, upon collapse, approaches to, reference conditions while the effects of the anthropogenic stressor persisted in the ponds (Figure 2). The amplitudes of fluctuation followed no regular patterns, which suggested a new cause-and-effect mechanism for disturbance ecology described as a "protracted press disturbance – roller coaster response" relationship (Angeler and Moreno in press). This model emphasizes stochastic oscillations in community composition, punctuated by periods when the community approaches reference conditions. From the applied viewpoint, this model suggests that the accurate detection of perturbation and the implementation of sound management and restoration strategies will require intensive sampling designs that span multiple hydroperiods in persistently degraded ponds.

Submerged macrophyte-dominated state   Phytoplankton-dominated state

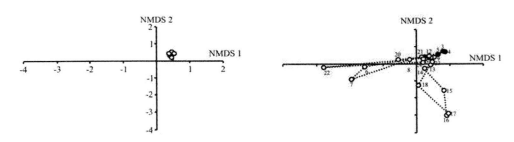

Figure 2. Non-metric multidimensional scaling plots showing zooplankton community similarity and temporal trends in the submerged macrophyte-dominated state and the phytoplankton-dominated state of temporary ponds. The state shift was induced by contamination with a fire retardant chemical (Fire Trol 934). Black and white dots show the data from the pre-contamination and post-contamination period, respectively, and the numbers above the dots indicate time trajectories in months. Note that sampling points in the submerged macrophyte-dominated state show a high overlap in ordination space throughout the study in comparison with the ordination shown for the phytoplankton-dominated state. This suggests that fluctuation amplitudes are markedly increased as a result of anthropogenic stress. The figure is modified from Angeler and Moreno (in press).

2. Local and Regional Impacts in Las Tablas De Daimiel Floodplain Wetland (Spain)

Las Tablas de Daimiel National Park (TDNP) is a floodplain wetland, located in semi-arid Central Spain. The wetland was once a unique landscape in Europe, in terms of complex hydrological functioning with combined surface flooding of two rivers and groundwater discharge, and biological characteristics that sustained local societies (Alvarez-Cobelas and Cirujano 1996). It provides an example where cumulative anthropogenic impacts at the local and regional level contributed to large-scale ecosystem shifts (Alvarez-Cobelas et al. 2001). These impacts included drainage of 130 km^2 which disrupted surface flooding in the 1960s and early 1970s. Intensive agricultural irrigation in the catchment during the second half of the 20th century disconnected the aquifer from the surface wetland and permanently severed inputs from the Guadiana River, the main inflow at the eastern part of TDNP. As a result of increased dryness, this site suffered from burning peat which caused the terrain to subside, turning the original inflow channel into an outflow point instead. In the 1980s, the wetland also served adjacent villages as a wastewater discharge area which converted TDNP into a hypereutrophic ecosystem.

To prevent further degradation, TDNP was declared a National Park in 1973, and a terminal dam was built in 1988 to retain water in the wetland. The latter is an example where short-sighted hydrological management had far-reaching effects on the ecological integrity of the system. This dam induced a catastrophic state shift that altered essentially all abiotic (hydrology [Sánchez-Carrillo 2000], sedimentation [Sánchez-Carrillo et al. 2001], nutrient dynamics [Sánchez-Carrillo and Alvarez-Cobelas, 2001]) and biotic components, especially emergent vegetation (Alvarez-Cobelas et al. in review). As a result, TDNP has become locked in a novel hypereutrophic state (Alvarez-Cobelas et al. in press) without the possibility of being restored to pristine conditions given its anthropogenic stress history. Unfortunately, sound management of this "novel" ecosystem is currently also limited because of conflicting interest from irrigation parties and farmer collectives.

3. Secondary Salinisation in South-Western Australian Wetlands

Another serious impact of land clearing in semi-arid areas is secondary salinisation of temporary wetlands. Davis et al. (2003) used the paradigm of multiple stable states to characterise the salinity-driven ecological changes occurring in salinised wetlands of south-west Australia. Four different equilibria were viewed as alternative ecological regimes: (i) clear water dominated by submerged aquatic macrophytes, (ii) clear water dominated by benthic microbial communities (BMCs), (iii) turbid water dominated by phytoplankton, and (iv) turbid water dominated by sediment. Increased salinity was predicted to cause (i) to shift to (ii) whereas increased nutrient loading would drive (i) to (iii) and (ii) to (iii) (Davis et al. 2003). The management goal for salinised temporary wetlands of south-western Australia is (i) as it is considered most similar to the historic natural state and performs a more diverse range of ecological functions (Sim et al. 2006a).

Transitions between macrophyte-dominated and BMC-dominated states may not necessarily be evidence for ASSs because the regular drying patterns seemed to prevent the appearance of strong feed-back mechanisms that might maintain the BMC-dominated state (Sim et al. 2006b). Regime change in these temporary wetlands was driven by the combined effects of salinity and water regime on species life histories and their competitive ability. For example, salinity constrained the degree to which a macrophyte community could establish and persist (Sim et al. 2006a), enabling BMC-dominated states to persist at high salinities. When salinity was low enough for macrophytes to establish, regular drying interspersed with hydroperiods of at least 4-5 months allowed macrophytes to dominate, regardless of the previous state (Sim et al. 2006b). Nonetheless, knowledge of the relationships between these states and the factors that drive their transitions is essential for effective management towards the goal of (i) but it does caution that ASS may not be an exclusively appropriate model for all temporary wetlands.

4. Drying up of Temporary Floodplains (Australia)

Many dryland rivers have extensive floodplains that support a variety of wetlands that connect to the main river during irregular floods. However, the intervals between floods of many of these temporary floodplain wetlands are increasing due to anthropogenic water extractions (Kingsford 2000b). When the floodplain is inundated, resting stages of many groups of invertebrates are triggered to hatch, providing a 'plankton soup' that fuels higher trophic levels such as fish and filter-feeding water birds. The dryland floodplain wetlands thus change from a state of dormancy ('bust') to one of flourishing biological productivity ('boom') in a dramatic sequence that often culminates in massive waterbird breeding events (e.g., Kingsford and Auld 2005).

Experimental microcosm studies inundating sediments from sites with different times since last flood have found that microinvertebrate densities and diversity decline as time since flooding increases (Boulton and Lloyd 1992, Jenkins and Boulton 1998) with especially sharp declines in output when dry periods exceed about 7-10 years. The implications of this for management are obvious (Jenkins and Boulton in press) because it seems that time since last flood may be a strong driver in the ASSs of many dryland river floodplains. Therefore, knowing the primary driver for this transition allows river managers to allocate environmental flows at suitable periods to promote occasional 'boom' states. The success of such manipulations could be measured using resting stages of floodplain invertebrates (Jenkins and Boulton, in press) as well as the more commonly-used waterbird breeding events (Kingsford and Auld 2005). There is an exciting array of future research opportunities in these temporary floodplain wetlands to explore how other variables known to change with time since last flooding (e.g., organic matter content and quality, available nutrients, metabolism, K. Jenkins unpublished data) might affect the rates of change and persistence of different states. As time since flooding increases due to increased water use and loss of flooding, will these temporary floodplain wetlands be locked into new stable states? If so, how persistent might these states be? With environmental water allocations, will their assemblages recover to the initial state? Although microcosm experiments hold promise for testing many of these predictions, floodplain-scale field studies are crucial to deal with such 'open systems' and there is great scope for adaptive management of responses to environmental water allocations to floodplain wetlands (see later).

5. Impounding Temporary Wetlands for Water Storage

The converse of the anthropogenically increased duration of drying described above is when temporary floodplain wetlands become permanently inundated when they are used to store water. Often, there is a perception that drying is a 'negative' process and that a management goal in arid areas should be the provision of more bodies of permanent water

(Boulton et al. 2000). There are two potential problems with these novel 'constructed' states. The first relates to the loss of natural hydrological variability (including drying) that seems to underpin nearly all ecological and biogeochemical processes in these types of wetlands. For example, a 19-year study of waterbirds across 12 floodplain wetlands showed that permanent inundation of six of these wetlands shifted them to an alternative state characterised by lower densities and diversity of waterbirds than observed in the other six temporary wetlands (Kingsford et al. 2004). Reduced hydrological variability also impacts on virtually every other biotic component of these temporary wetlands (reviewed in Jenkins et al. 2005).

The second issue in these constructed states relates to the way that the new hydrological conditions favour other impacts such as exotic fish and their persistence in the system. It is unclear whether alternate states of temporary waters that have been used as storages and then decommissioned and allowed to return to their previous state are irreversibly changed when exotic species such as carp have been present for long periods. Encouragingly, waterbird densities and diversity did rise sharply when a water storage lake dried (Kingsford et al. 2004). Nevertheless, 'secondary impacts' may complicate the use of ASS models for predicting the return to a preferred state by simply manipulating hydrology, and again, is a rich area for research and adaptive management.

ASSs, MANAGEMENT, AND FUTURE RESEARCH

These case studies provide examples of different, context-dependent forms of anthropogenic impacts that affect temporary wetlands. Most of these studies are based on non-manipulative field studies and small-scale experiments which inherently limit unambiguous demonstration, interpretation and explanation of ASSs (Schröder et al. 2005). Even manipulative field experiments may not be the ultimate proof of the existence of ASSs if carried out over only short spatiotemporal scales. For example, case study 1 reported on state shifts in temporary ponds as a result of fire retardant contamination that were consistent with observations made in shallow lakes. Have ponds really been nudged to an alternative state or has the perturbation induced transient dynamics from which the recovery time is longer than two hydroperiods?

This example highlights a hotly debated issue in current ecology. Some authors suggest that ecosystems that are strongly regulated by abiotic forces readily shift between ASSs (Didham et al. 2005). Others defend the idea that systems with high perturbation frequencies will never reach a presumable end state regardless of how many there might be so that permanent transient dynamics become more important in driving ecosystem evolution. However, these views are not mutually exclusive and the review by Schröder and coworkers (2005) reconciles both arguments, suggesting that ASSs can still occur in permanently transient systems.

Unfortunately, long-term data, such as the waterbird data in case study 5 that could reveal strict ASSs or transient dynamic behaviour of temporary wetlands are scant. Paleolimnological reconstructions, including approaches that combine genetic/evolutionary and experimental studies in the context of resurrection ecology (Kerfoot et al. 1999) hold promise to address this topic, and this tool seems especially useful for studying aquatic ecosystem responses to global climate change (Angeler 2007). Insights gained from resurrection ecology can then be used with information about more recent ASSs to set management goals for temporary waters.

Transient Dynamics or Stable Equilibria: Implications for Management

Types, magnitudes and spatiotemporal patterns of anthropogenic stress combined with an ecosystem's capacity to tolerate impact, based on the life history traits of its constituent species, will mediate the response of ecological systems to perturbation. We will discuss this issue using two different scenarios and highlight consequences for management.

Scenario 1: Irreversible Damage to Temporary Waters

Some forms of water resource management (damming, channelisation) are 'sledge-hammer' stressors that produce a major state shift in many temporary wetlands. These create new self-reinforcing feedback loops (case study 5), which often act in concert with other stressors (e.g., contamination, exotic species as in case study 2 or salinity as in case study 3). Consequently, restoration of pristine conditions may be impossible unless these anthropogenic stressors are removed. Removing stressors may sometimes not be enough to bring a system back to its pristine state and may require additional efforts to break equilibrium conditions that build in degraded states (see below). Major challenges lie in managing the resulting novel ecosystems in ways to maintain their ecological sustainability. Enhancing the resilience of critical functional groups to future disturbance, rather than restoring rare species, could be crucial for this task (Bellwood et al. 2004).

Scenario 2: Reversible Damage to Temporary Waters

Other forms of anthropogenic stress have the potential to move aquatic ecosystems to an alternative equilibrium, yet the self-reinforcing feedback mechanisms can be disrupted to allow a return to a pristine state. As is known from shallow lakes, recovery pathways toward restoration of macrophyte-dominance follow distinct trajectories compared to those that take place during degradation (Suding and Gross 2006). For example, restoration of shallow lakes requires substantially more efforts (e.g., control of nutrient loading, food web manipulations, sediment consolidation) than controlling nutrients alone. Revisiting case study 1, management and restoration of retardant contaminated ponds will be more challenging in the context of ASS theory compared with transient dynamics theory. For example, imagine toxicity effects on degraded pond seed banks that limit recolonisation of plants in the long term (evidence for a hysteresis effect). In this case, reinstating the macrophyte-dominated state will likely require sediment dredging and the application of donor seed banks from other wetlands. In contrast, if submerged macrophyte growth fails temporarily simply because of poor water quality (cf. case study 3), a "sit and wait" management strategy to let the water clear again, enabling the plants to re-establish may be sufficient (and more cost effective). Clearly, because ecosystem processes and communities of temporary wetlands are strongly regulated by hydrological fluctuation (Mitsch and Gosselink 2000), interactions of natural disturbance regimes with species life history traits and their competitive ability (shown in case study 3) will determine transient processes or abrupt state shifts in response to anthropogenic stress.

The Role of Food Webs in Temporary Wetlands

Wetland management has traditionally focused on restoring the original abiotic features of the system, whilst ignoring changes in biotic factors and the feedback between biotic and abiotic factors that have developed in the degraded state (Zedler 2000; Suding et al. 2004; Suding and Gross 2006). Food webs have not been studied as exhaustively in temporary waters as in deep and shallow lakes, and the influence of landscape features on local processes in temporary wetlands is almost unknown. Whether physico-chemical conditions, water regime or other factors are driving food web structures or whether food web structure drives other changes in wetland ecosystems is therefore poorly understood, hampering effective management.

Temporary wetland food webs are highly variable in space and time and depend on life-history traits of their constituent species, and characteristics of adjacent terrestrial environments (Angeler and Alvarez-Cobelas 2005). Temporary wetland biota can be generally classified into two broad categories: 1) organisms with poor overland dispersal capacity which survive dry periods in form of resting stages in sediments (resident species), and 2) organisms that lack dormancy in their life cycle but which can avoid dry periods by actively seeking suitable habitat (migratory species). Temporary wetland food webs typically respond to bottom-up regulation when wetlands fill after rainfall events and recruitment from sediments takes place. With increasing length of hydroperiod, many wetlands become colonised by mobile predators which exert a strong top-down control on the lower trophic levels (Wellborn et al. 1996). Emergence from wetland soil seed banks has been proposed recently as a useful tool to asses ecological integrity or the impact of anthropogenic stress in wetlands (Angeler and García 2005), and some of our case studies show a reduced seed bank potential as a result of degradation (case study 4). Can such alterations cascade up the food web to alter the colonisation sequence of predators and hence, the strength of top-down control in temporary waters? Perhaps detecting declines in dormant biota can provide early warnings of impending impacts on higher trophic levels through loss of their food supply (Jenkins and Boulton in press).

The spatial distribution of wetlands is also critical. For example, habitat fragmentation may affect the nature of top-down control because when the geographic isolation between wetlands increases, the active dispersal of predators becomes limited. As wetlands lie along gradients of habitat permanence and geographical isolation, we hypothesise that the relative contributions of bottom-up and top-down regulation of temporary wetland communities and ecosystem processes will be context-specific. Strong site idiosyncrasies, based on local and regional settings, suggest that current models of ASSs in aquatic ecology have been too simplistic. Integrative studies based on landscape ecology, metacommunity dynamics and island biogeography theory could help determine whether a system is controlled from the bottom up or top down, whether landscape features mediate in these processes, and whether the type of control generates mechanisms that helps maintain a system in a given ASS (Navarrete and Berlow 2006). This knowledge would be invaluable to inform effective wetland management and restoration (Vander Zanden et al. 2006), especially at a landscape scale rather than the all-too-frequent piece-meal approach at the scale of individual wetlands.

Temporary Wetlands, Global Climate Change and the Future

Temporary wetlands are among the ecosystems most threatened by global climate change. Global warming likely decreases the length of hydroperiods or reduces the frequencies of flooding events (Alvarez-Cobelas et al. 2005) thereby altering the natural hydrological functioning of temporary wetlands. Furthermore, the projected increase in human populations in the next century will increase the demand for drinking water supply and agricultural irrigation, which could lead to a series of ecosystem alterations which will profoundly alter temporary wetlands. Changes in land-use practices, spread of exotic species, pollution events, and habitat loss and fragmentation pose additional threats. All these stressors are likely to interact in strongly non-linear, complex and discontinuous ways (IPCC 2001), confounding prediction of the novel ASSs. Adaptive management of increasingly threatened temporary wetlands will pose challenges to the survival of their unique biota and management of the ecosystem services they provide but we suggest that the ASS paradigm may provide a useful framework based on the findings from the case studies presented above.

REFERENCES

Alvarez-Cobelas, M. and Cirujano, S. (eds). 1996. Las Tablas de Daimiel: *Ecología acuática y Sociedad*. Ministry of Environment, Madrid.

Alvarez-Cobelas, M., Cirujano, S. and Sánchez-Carrillo, S. 2001. Hydrological and botanical man-made changes in the Spanish wetland of Las Tablas de Daimiel. *Biological Conservation* 97: 89-98.

Alvarez-Cobelas, M., Catalan, J., and García de Jalon, D. 2005. Impacts on inland aquatic ecosystems. *In* J.M. Moreno (Ed.) *Effects of climate change in Spain*, pp. 113-146, Ministerio de Medio Ambiente, Madrid.

Alvarez-Cobelas, M., Sánchez-Carrillo, S. and Cirujano, S. 2007. Strong site effects dictate nutrient patterns in a Mediterranean floodplain. *Wetlands* 27 (in press)

Alvarez-Cobelas, M., Cirujano, S., Sánchez-Carrillo, S. and Angeler, D.G. In review. Long-term changes in spatial patterns of emergent vegetation in a Mediterranean floodplain: natural vs anthropogenic constraints. *Plant Ecology.*

Angeler, D.G. 2007. Resurrection ecology and global climate change research in freshwater ecosystems. *Journal of the North American Benthological Society* 26: 12-22.

Angeler, D.G., and Alvarez-Cobelas, M. 2005. Island biogeography and landscape structure: integrating ecological concepts in a landscape perspective of anthropogenic impacts in temporary wetlands. *Environmental Polution.* 138: 421–425.

Angeler, D.G., and García, G. 2005. Using emergence from soil propagule banks as indicators of ecological integrity in wetlands: advantages and limitations. *Journal of the North American Benthological Society* 24: 740– 752.

Angeler, D.G. and Moreno, J.M. 2006. Impact-recovery patterns of water quality in temporary wetlands after fire retardant contamination. *Canadian Journal of Fisheries and Aquatic Sciences* 63: 1617-1626.

Angeler, D.G. and Moreno, J.M. in press. Zooplankton community recovery after press-type anthropogenic stress in temporary ponds. *Ecological Applications*

Beisner, B.E., Haydon, D.T. and Cuddington, K. 2003. Alternative stable states in ecology. *Frontiers in Ecology and the Environment* 1: 376–382.

Bellwood, D.R., Hughes, T.P., Folke, C. et al. 2004. Confronting the coral reef crisis. *Nature* 429: 827-833.

Boulton, A.J. and Lloyd, L.N. 1992. Flooding frequency and invertebrate emergence from dry floodplain sediments of the River Murray, Australia. *Regulated Rivers: Research and Management* 7: 137-151.

Boulton, A.J., Sheldon, F., Thoms, M.C. and Stanley, E.H. 2000. Problems and constraints in managing rivers with variable flow regimes. *In* P.J. Boon, B.R. Davies and G.E. Petts (Eds) *Global Perspectives on River Conservation: Science, Policy and Practice*, pp. 415-430, John Wiley and Sons, London.

Chase, J.M. 2003a. Experimental evidence for alternative stable equilibria in a benthic pond food web. *Ecology Letters* 6: 733–741.

Chase, J.M. 2003b. Community assembly: when should history matter? *Oecologia* 136: 489–498.

Cottenie, K., Nuytten, N., Michels, E. et al. 2001. Zooplankton community structure and environmental conditions in a set of interconnected ponds. *Hydrobiologia* 442: 339- 350.

Comín, F.A. and Williams, W.D. 1994. Parched continents: Our common future? *In* R. Margalef (Ed.), *Limnology Now: A Paradigm of Planetary Problems*. pp. 473-527, Elsevier Science, BV.

Davis J., McGuire M., Halse S., Hamilton D., Horwitz P., McComb A., Froend R., Lyons M. and Sim L. 2003. What happens when you add salt: predicting impacts of secondary salinisation on shallow aquatic ecosystems using an alternative states model. *Australian Journal of Botany* 51: 715–724.

DeClerck, S., De Bie, T., Ercken, D., Hampel, H., Schrijvers, S., VanWichelen, J., Gillard, V. et al. 2006. Ecological characteristics of small farmland ponds: Associations with land use practices at multiple spatial scales. *Biological Conservation* 131: 523-532.

Dent, C.L., Cumming, G.S. and Carpenter, S.R. 2002. Multiple states in river and lake ecosystems. *Philosophical Transactions of the Royal Society, Series B* 357: 635–645.

Didham, R.K., Watts C.H. and Norton D.A. 2005. Are systems with strong underlying abiotic regimes more likely to exhibit alternative stable states? *Oikos* 110: 409-416.

Falk, D.A., Palmer, M.A., and Zedler, J. B. (eds.) 2006. *Foundations of Restoration Ecology*. Island Press, Washington, U.S.A.

Gibbs, J. P. 2000. Wetland loss and biodiversity conservation. *Conservation Biology* 14: 314–317.

IPCC (Intergovernmental Panel of Climate Change). 2001. Climate change 2001: synthesis report. Volume 4 *in* R.T. Watson (Ed.), *Third Assessment Report of the Intergovernmental Panel of Climate Change (IPCC)*. Cambridge University Press, Cambridge, UK.

Jenkins, K.M. and Boulton, A.J. 1998. Community dynamics of invertebrates emerging from reflooded lake sediments: flood pulse and aeolian influences. *International Journal of Ecology and Environmental Science* 24: 179-192.

Jenkins, K.M. and Boulton, A.J. In press. Detecting impacts and setting restoration targets in arid-zone rivers: aquatic microinvertebrate responses to loss of floodplain inundation. *Journal of Applied Ecology*.

Jenkins, K.M., Boulton, A.J. and Ryder, D.S. 2005. A common parched future? Research and management of Australian arid-zone floodplain wetlands. *Hydrobiologia* 552: 57-73.

Jeppesen, E., Jensen, J. P., Søndergaard, M., Lauridsen, T., Pedersen, J. L. and Jensen, L. 1997. Top-down control in freshwater lakes: The role of nutrient state, submerged macrophytes and water depth. *Hydrobiologia* 342/343:151-164.

Kerfoot, W.C., Robins, J.A., and Weider, L.J. 1999. A new approach to historical reconstruction: combining descriptive and experimental paleolimnology. *Limnology and Oceanography* 44: 1232-1247.

Kingsford, R.T. (2000a) Ecological impacts of dams, water diversions and river management on floodplain wetlands in Australia. *Austral Ecology*, 25(2), 109-27.

Kingsford RT 2000b. Protecting rivers in arid regions or pumping them dry? *Hydrobiologia* 427: 1-11.

Kingsford RT, Auld KM 2005. Waterbird breeding and environmental flow management in the Macquarie Marshes, arid Australia. *River Research and Applications* 21: 187-200.

Kingsford R.T., Jenkins K.M. and Porter J.L. 2004. Imposed hydrological stability on lakes in arid Australia and effects on waterbirds. *Ecology* 85: 2478-2492.

Lewontin, R. 1969. The meaning of stability. *In* G.M. Woodwell and H.H. Smith (Eds) *Diversity and Stability in Ecological Systems.* pp. 13–24. Brookhaven Symposium Biology 22. Brookhaven National Laboratory, Brookhaven, New York.

Mitsch, W. J. and Gosselink, J. G. 2000. *Wetlands*, 3rd ed. Wiley and Sons, New York.

Navarrete, S.A. and Berlow, E.L. (2006) Variable interaction strengths stabilize marine community pattern. *Ecology Letters* 9, 526-536.

Robinson, A. 1995. Small and seasonal does not mean insignificant: why it's worth standing up for tiny and temporary wetlands. *Journal of Soil and Water Conservation* 50: 586-590.

Ruggiero, A., Solomini, A.G., and Carachini, G. 2005. The alternative stable state concept and the management of Apennine mountain ponds. *Aquatic Conservation: Marine and Freshwater Ecosystems* 15: 625–634.

Sánchez-Carrillo, S. 2000. *Hidrología y sedimentación actual de Las Tablas de Daimiel*. Ph. D. Thesis. Universidad Autónoma, Madrid.

Sánchez-Carrillo, S. and Alvarez-Cobelas, M. 2001. Nutrient dynamics and eutrophication patterns in a semiarid wetland: the effects of fluctuating hydrology. *Water, Air and Soil Pollution* 131: 97-118.

Sánchez-Carrillo, S., Alvarez-Cobelas, M., and Angeler, D.G. 2001. Sedimentation in the semiarid freshwater wetland Las Tablas de Daimiel (Central Spain). *Wetlands* 21: 112–124.

Scheffer, M., Hosper, S. H., Meijer, M.-L., Moss, B. and Jeppesen, E. (1993) Alternative equilibria in shallow lakes. *Trends in Ecology and Evolution* 8: 275-279.

Scheffer, M., Carpenter, S.R., Foley, J.A. et al. (2001) Catastrophic shifts in ecosystems. *Nature* 413: 591–96.

Schröder, A., Persson, L. and De Roos, A.M. 2005. Direct experimental evidence for alternative stable states: a review. *Oikos* 110: 3-19.

Sim LL, Chambers JM, Davis JA 2006a. Ecological regime shifts in salinised wetland systems. I. Salinity thresholds for the loss of submerged macrophytes. *Hydrobiologia* 573:89-107.

Sim, L.L., Davis, J.A., Chambers, J.M. and Strehlow, K. 2006b. What evidence exists for alternative ecological regimes in salinising wetlands? *Freshwater Biology* 51: 1229–1248

Suding, K.N., Gross, K.L. and Houseman, G.R. 2004. Alternative states and positive feedbacks in restoration ecology. *Trends in Ecology and Evolution* 19: 46-53.

Suding, K.N. and Gross, K.L. (2006). The dynamic nature of ecological systems: multiple states and restoration trajectories. *In* D.A. Falk, M.A. Palmer and J.B. Zedler (Eds.), *Foundations of Restoration Ecology*, pp. 190 – 209. Island Press, Washington, D.C.

Vander Zanden, M.J., Olden, J.D. and Gratton, C. (2006) Food-web approaches in restoration ecology. *In* D.A. Falk, M.A. Palmer and J.B. Zedler (Eds.), *Foundations of Restoration Ecology*, pp. 165-189. Island Press, Washington, D.C.

Wellborn, G.A., Skelly, D. K., and Werner, E. E. (1996) Mechanisms creating community structure across a freshwater habitat gradient. *Annual Reviews of Ecology and Systematics* 27, 337–363.

Williams, D.D. (2006) *The Biology of Temporary Waters*. Oxford University Press, Oxford, U.K.

Zedler, J.B. (2000) Progress in wetland restoration ecology. *Trends in Ecology and Evolution* 15: 402-407.

In: Environmental Research Advances
Editor: Peter A. Clarkson, pp. 19-43

ISBN: 978-1-60021-762-3
© 2007 Nova Science Publishers, Inc.

Chapter 1

NATURAL RESOURCES
WOODY BIOMASS USERS' EXPERIENCES
OFFER INSIGHTS FOR GOVERNMENT
EFFORTS AIMED AT PROMOTING ITS USE[*]

United States Government Accountability Office

ABSTRACT

Financial incentives and benefits associated with using woody biomass were the primary factors facilitating its use among the 13 users GAO reviewed. Four users received financial assistance (such as state or federal grants) to begin their use of woody biomass, three received ongoing financial support related to its use, and several reported energy cost savings over fossil fuels. Using woody biomass also was attractive to some users because it was available, affordable, and environmentally beneficial.

Several users GAO reviewed, however, cited challenges in using woody biomass, such as difficulty obtaining a sufficient supply of the material. For example, two power plants reported running at about 60 percent of capacity because they could not obtain enough material. Some users also reported that they had difficulty obtaining woody biomass from federal lands, instead relying on woody biomass from private lands or on alternatives such as sawmill residues. Some users also cited increased equipment and maintenance costs associated with using the material.

The experiences of the 13 users offer several important insights for the federal government to consider as it attempts to promote greater use of woody biomass. First, if not appropriately designed, efforts to encourage its use may simply stimulate the use of sawmill residues or other alternative wood materials, which some users stated are cheaper or easier to use than woody biomass. Second, the lack of a local logging and milling infrastructure to collect and process forest materials may limit the availability of woody biomass; thus, government activities may be more effective in stimulating its use if they take into account the extent of infrastructure in place. Similarly, government activities such as awarding grants or supplying woody biomass may stimulate its use more effectively if they are tailored to the scale and nature of the targeted users. However,

---

[*] Excerpted from GAO Report 07-464T, dated February 6, 2007.

agencies must remain alert to potential unintended ecological consequences of their efforts.

EXAMPLES OF WOODY BIOMASS USERS GAO REVIEWED

Pulp and paper mill.   Wood-fired heating facility at rural school.

In recent years, extensive wildland fires have drawn attention to the abnormally dense vegetation in many of our nation's forests. The federal government has responded by placing greater emphasis on reducing the danger of such fires by thinning forests and rangelands to help reduce the buildup of potentially hazardous fuels. These thinning efforts are expected to generate large amounts of woody material, including many small trees, limbs, and brush—often referred to as woody biomass—that traditionally have had little commercial value [1].

Widespread thinning efforts will be costly to the federal government. To help defray these costs, and to enhance rural employment and economic development, the government is promoting a market for woody biomass. However, as we have reported in the past, [2] the increased use of woody biomass faces several obstacles. Officials in federal agencies seeking to promote its use—including the Departments of Agriculture, Energy, and the Interior—told us that woody biomass use is hampered by the high costs of removing and transporting it from forests and the difficulty in obtaining a reliable supply in some areas. Nevertheless, a number of businesses and government entities are using woody biomass for various purposes, including heating schools and hospitals, making lumber and other products, and generating electricity.

In this context, and in response to our previous report describing agency activities to promote woody biomass, you asked us to review current users of woody biomass to determine whether their experiences offer any insights for expanding its use. Specifically, we agreed to (1) identify key factors facilitating the use of woody biomass among selected users, (2) identify challenges these users have faced in using woody biomass, and (3) discuss any insights that our findings may offer for promoting greater use of woody biomass.

To conduct our review, we used a structured interview guide to collect information from 13 users of woody biomass, including power plants, pulp and paper mills, and school and hospital facilities in various locations around the United States. Appendix I contains information about each of the 13 users in our review. We first identified users by interviewing federal and nonfederal officials knowledgeable about the use of woody biomass and by reviewing pertinent documents such as federal agency studies of woody biomass utilization. Users in our review were then selected from a range of industries and geographic regions. The information we collected about these 13 entities should not be generalized to other woody biomass users because of variations in the characteristics of different users. Appendix II provides further details on the scope and methodology of our review. We conducted our work from May 2005 through January 2006 in accordance with generally accepted government auditing standards.

RESULTS IN BRIEF

The primary factors facilitating woody biomass use among users we reviewed were financial incentives or benefits associated with its use, either in the form of financial assistance for using the material or in energy cost savings. Other factors included the availability of an affordable supply of woody biomass and users' interest in environmental benefits associated with its use. Four of the 13 users in our review received financial assistance to begin using woody biomass, including state and federal grants and tax-exempt bonds. Three users also were given ongoing support as a result of their use of woody biomass, including grant funds for expanding their wood storage facilities and payments for producing renewable energy. Moreover, six users reported energy cost savings from using woody biomass in place of fossil fuels such as natural gas. For example, two small school districts individually reported about $50,000 and $60,000 in annual fuel cost savings, while two large pulp and paper mills reported several million dollars in such savings. Several of the 13 users also cited the availability of an affordable supply of the material as important in their use of woody biomass—particularly in cases where it was already being removed as a byproduct of other activities, such as commercial logging or private land clearing. Finally, three users told us that their use of woody biomass was due in part to anticipated environmental benefits associated with using the material, including improved forest health and reduced emissions.

Using woody biomass, however, was not without challenges for the users we reviewed. Users cited insufficient supply, increased equipment and maintenance costs, and other factors that limited their use of woody biomass or made it more difficult or expensive to use. In contrast to users citing an available supply of woody biomass, seven users reported they found it difficult or impossible to obtain a sufficient supply of the material. For example, two power plants reported running at about 60 percent of their capacity because they could not obtain enough material to operate at full capacity. Five users told us they had difficulty obtaining woody biomass from federal lands, which was of particular concern to users located in areas where federal lands constitute a substantial portion of the landscape. Such users relied more on woody biomass from private lands or on alternative wood materials such as sawmill residues (including sawdust, chips, bark, and similar materials) or urban wood waste (made up of tree trimmings, construction debris, and the like). Several users also told us that, despite the financial advantages of using woody biomass in place of oil or natural gas, they had incurred costs in using woody biomass that they would not have incurred had they burned these other fuels. Users cited costs for additional wood-handling equipment, such as storage bins and conveyors, and added operation and maintenance costs, including costs arising from problems in storing and handling woody material.

Our findings offer several insights for promoting greater use of woody biomass, specifically: (1) attempts to encourage the use of woody biomass may serve to stimulate the use of alternative wood materials such as sawmill residues instead, (2) government activities may be more effective in stimulating woody biomass use if they take into account the extent of logging and milling infrastructure, and (3) efforts to encourage woody biomass use may need to be tailored to the scale and nature of individual recipients' use.

- If not appropriately designed, attempts to encourage the use of woody biomass may simply stimulate the use of mill residues or other alternative wood materials, which some users told us are cheaper or easier to use than woody biomass. For example, in 2003, the Forest Service provided a grant to fund a Montana school's conversion to a wood heating system in order to stimulate the use of woody biomass in the area. However, at the time of our review, the school was using less expensive wood residues from a nearby log-home builder rather than woody biomass. Further, in using woody biomass, users in our review often used the tops and limbs from trees harvested for merchantable timber or other uses rather than the small-diameter trees that contribute to the problem of overstocked forests. As the federal government seeks to stimulate the market for materials that result from forest-thinning activities, it should consider the potential impacts of its actions to ensure that they promote greater use of small-diameter trees and not simply increase the use of other wood materials.

- Government activities may be more effective in stimulating woody biomass use if they take into account the extent to which a logging and milling infrastructure is in place in potential users' locations. The availability of a reasonably low-cost supply of woody biomass depends in part on the presence of a local logging and milling infrastructure to collect and process forest materials, even though this infrastructure also generates alternatives to woody biomass. Without a milling infrastructure, there may be little demand for forest materials, and without a logging infrastructure, there may be no way to obtain the materials. Indeed, officials at one power plant operating at a reduced capacity because of a shortage of wood for the plant told us that the shortage was due to the lack of a local logging infrastructure—in other words, there simply were not enough loggers to carry out needed forest projects, and it was not cost-effective for the plant to obtain material from more distant sources. In general, the type and amount of effort needed to increase the use of woody biomass may vary among locations, depending on the extent to which a logging and processing infrastructure is already in place. The presence of such an infrastructure, however, may also increase the availability of mill residues—potentially complicating efforts to promote woody biomass use by offering cheaper or more readily available alternative materials.

- Similarly, government activities may be more effective in stimulating woody biomass use if their efforts are tailored to the scale and nature of the users being targeted. Most of the large wood users we reviewed, such as pulp and paper mills or power plants, were primarily concerned about supply, and thus might benefit most from federal efforts to provide a predictable and stable supply of woody biomass. In fact, one company currently plans to build a woody biomass power plant in eastern Arizona largely in response to a nearby federal thinning project that is expected to last 10 years and generate a stable, long-term supply of the material. In contrast, small users we reviewed did not express concerns about the availability of supply, in part because their consumption was relatively small; however, several relied on external financing for their up-front costs to convert to woody biomass use. Such users might benefit most from financial assistance such as grants or loan guarantees to fund initial conversion efforts, and indeed, federal agencies are providing grants intended to promote the use of woody biomass, including a Forest Service grant program specifically intended to help defray federal thinning costs by stimulating woody biomass use. However, agencies must remain alert to potential unintended consequences of their efforts to stimulate the use of woody biomass. As we noted in our prior report, some officials expressed concern that developing a market for woody biomass could result in adverse ecological consequences such as unnecessary forest thinning to meet demand for the material. Further, while agency grants to woody biomass users may provide the users with benefits such as fuel cost savings, these grants may not in all cases defray agency thinning costs.

In responding to a draft of this report, the Departments of Agriculture, Energy, and the Interior all generally agreed with our findings.

BACKGROUND

Woody biomass—small-diameter trees, branches, and the like—is generated as a result of timber-related activities in forests or rangelands. Small-diameter trees may be removed to reduce the risk of wildland fire or to improve forest health, while treetops, branches, and limbs, collectively known as "slash," are often the byproduct of traditional logging activities or thinning projects. Slash is generally removed from trees on site, before the logs are hauled for processing. It may be scattered on the ground and left to decay or to burn in a subsequent prescribed fire, or piled and either burned or hauled away for use or disposal. Figure 1 depicts woody biomass in the form of small-diameter logs and slash.

Woody biomass, both small-diameter logs and slash, can be put to various uses. Small-diameter logs can be sawed into structural lumber, particularly as some sawmills have retooled to process these logs in addition to, or instead of, larger logs. Other users of whole small-diameter logs include some log-home builders and post and pole makers. After bark, branches, and leaves are removed, logs can be chipped and processed to make pulp, the raw material from which paper, cardboard, and other products are made. Chipped wood also is used by manufacturers of oriented strand board and other such engineered wood products. Both small-diameter logs and slash also can be chipped or ground and used for fuel, either in raw form or after being dried and made into fuel pellets. Various entities, including power plants, schools, pulp and paper mills, and others, burn woody biomass in boilers to turn water into steam, which is used to make electricity, heat or cool buildings, or provide heat for industrial processes.

Figure 1. Small-Diameter Logs and Slash Generated from a Montana Fuels Reduction Project.

Federal, state, and local governments, as well as private organizations, are working to expand the use of woody biomass. Recent federal legislation, including the Biomass Research and Development Act of 2000, [3] Healthy Forests Restoration Act of 2003, [4] Consolidated Appropriations Act for Fiscal Year 2005, [5] and Energy Policy Act of 2005, [6] contains provisions for woody biomass research and financial assistance. For example, the Consolidated Appropriations Act for Fiscal Year 2005 made up to $5 million in appropriations available for grants to create incentives for increased use of woody biomass from national forest lands; in response, the Forest Service awarded $4.4 million in such grants in fiscal year 2005. State and local governments also are encouraging the use of woody biomass through grants, research, and technical assistance. For example, the Bitter Root Resource Conservation and Development Council, a nonprofit organization sponsored by state government entities and three counties in Montana, [7] is helping to coordinate a federally funded effort—known as the Fuels for Schools program—to install wood-fired heating systems in rural school buildings. Other states, such as Idaho, Nevada, and North Dakota, also are participating in the Fuels for Schools program.

Private corporations also are researching new ways of using woody biomass and wood waste, often in partnership with government and universities. For example, one corporation has partnered with the University of Georgia, and has developed and plans to license biorefinery technology for making chemicals, agricultural fertilizer, and transportation fuels such as ethanol from woody biomass. Another private company has developed technology that it hopes will significantly increase the ethanol yield from any type of biomass, including woody biomass.

Financial Incentives and Benefits, Access to an Affordable Supply, and Environmental Benefits Facilitated the Use of Woody Biomass among Users We Reviewed

The users in our review cited several factors contributing to their use of woody biomass, primarily financial incentives and benefits but also other factors such as an affordable supply of woody biomass and environmental considerations. Financial incentives encouraging the use of woody biomass included financial assistance, while financial benefits included energy cost savings from using woody biomass in place of other fuels. In addition, some users had access to a readily available and affordable supply of woody biomass, particularly in areas where material was being removed as part of commercial activities such as logging. Other users told us that their use of woody biomass was due in part to environmental or other perceived benefits.

Financial Incentives and Benefits Encouraged the Use of Woody Biomass by Several Users

Financial incentives for, and benefits from, using woody biomass were the primary factors for its use among several users we reviewed. Four of the 13 users in our review told us that initial financial assistance in the form of grants or bonds allowed them to begin using woody biomass. Three public entities—a state college in Nebraska, a state hospital in Georgia, and a rural school district in Montana—received financial grants covering the initial cost of the equipment that they needed to begin using woody biomass. In the case of the state college, a state grant of about $1 million in 1989 covered the cost of installing two wood-fired boilers used to heat about 1 million square feet of campus building space, as well as an expansion to the college's central heating plant to house the new boilers and the requisite wood storage and handling system [8]. The college received a subsequent grant of about $100,000 in 2003 to help defray the costs of installing a chiller powered by woody biomass, which supplies cool air to campus buildings. The state hospital in Georgia received about $2.5 million in state funds during the early 1980s to pay for the purchase and installation of wood-handling equipment, and the Montana school district received about $900,000 in federal funds in 2003 for the same purpose [9]. The fourth user—a wood-fired power plant in California—received financial assistance in the form of tax-exempt state bonds to finance a portion of the plant's construction, part of a statewide effort to promote the use of biomass power plants and thereby reduce air pollution created by burning the material in the open.

Three users in our review also received additional financial assistance, including subsidies and other payments that helped them continue their use of woody biomass.

- The California wood-fired power plant received about $10 per megawatt hour from the state government during 2003 and 2004, according to a plant official [10]. This subsidy, which also was provided to other biomass-fueled electricity producers in the state, was paid for by a "public goods" surcharge on consumers' utility bills. The plant also benefited from an artificially high price received for electricity during its first 10 years of operation, a result of California's implementation of the federal Public Utility Regulatory Policies Act of 1978 [11]. The act—a response to the unstable energy climate of the late 1970s—required utilities to purchase electricity from certain facilities producing electricity from renewable sources, including woody biomass, at prices established by state regulators [12]. However, the initial prices established by California—based on expectations of sharply rising oil and natural gas prices—proved to substantially exceed market prices in some years, benefiting this power plant by increasing its profit margin.
- The Montana school district also continues to receive financial assistance through its participation in the Fuels for Schools program. For example, the Bitter Root Resource Conservation and Development Council paid for the installation of a 1,000-ton wood fuel storage facility at the school district, capable of storing over a year's

supply of fuel. The council also financed the up-front purchase of a year's supply of fuel for the district, which the district repays as it uses the fuel. This ongoing assistance helped the district obtain wood fuel for about $24 per ton during the 2005-2006 school year, in contrast to the $36 per ton it paid for woody biomass in the previous year. Moreover, when some of the school district's wood fuel supply decayed more rapidly than expected, the council also arranged for the Forest Service to provide higher-quality woody biomass from a nearby fuels reduction project at a price of $10 per ton. Figure 2 shows the 1,000-ton wood fuel storage facility.

- One Colorado power plant that generated electricity by firing woody biomass with coal received ongoing financial benefits for using woody biomass by selling renewable energy certificates. Renewable energy certificates (sometimes referred to as "green tags") represent the environmental benefits of renewable energy generation—that is, the benefits of displacing electricity generated from nonrenewable sources, such as fossil fuels, from the regional or national electric grid. The certificates are sold separately from the electricity with which they are associated. Certificates can be purchased by utilities seeking to meet state requirements for renewable energy generation or by other entities seeking to support the use of renewable energy sources, and their sale can serve as an additional source of revenue to power plants using such sources. The Colorado power plant in our review generated about 730 megawatt hours of electricity through its use of woody biomass, [13] and sold the associated renewable energy certificates to the Forest Service for $23 per megawatt hour, or about $17,000 in total. The Forest Service purchased the certificates in order to promote woody biomass use and to offset the power plant's costs for using woody biomass.

Energy cost savings also were a major incentive for using woody biomass among six of the users we reviewed. Of the four users that produce central heat with wood, two users—small rural school districts in Pennsylvania and Montana—told us that they individually had saved about $50,000 and $60,000 in annual fuel costs by using wood instead of natural gas or fuel oil. Officials at one of these districts told us that these savings represented the equivalent of one teacher's annual salary, stating "we could either burn fuel oil and watch that money go up the chimney, or burn wood and put the money toward education." Likewise, the state college in Nebraska, which uses woody biomass to heat and cool about 1 million square feet of space in several campus buildings, typically saves about $120,000 to $150,000 annually, while the Georgia state hospital reported saving at least $150,000 in 1999, the last year for which information was available.

Figure 2. Wood Fuel Storage Facility at a Montana School District.

Similarly, the two pulp and paper mills we reviewed each reported saving several million dollars annually by using wood rather than natural gas or fuel oil to generate steam heat for their processes; officials at one mill stated that the mill's operating costs would increase significantly without the savings generated by burning wood, making it difficult for the mill to remain competitive. Each of these users told us that they planned to continue their use of woody biomass because they anticipated continuing high fossil fuel prices.

An Affordable Supply Facilitated the Use of Woody Biomass

An affordable supply of woody biomass facilitated its use, especially in areas where commercial activities such as logging or land clearing generated woody biomass as a byproduct. For example, the Nebraska state college was able to purchase woody biomass for an affordable price because logging activities in the area made slash readily available. Logging companies harvested timber in the vicinity of the college, hauling the logs to sawmills and leaving their slash; the college paid only the cost to collect, chip, and transport the slash to the college for burning. One official told us that without the area's logging activity, the affordable supply of woody biomass used by the college would be severely jeopardized and the college would have to pay much higher prices to heat and cool its campus.

Two Pennsylvania users in our review also obtained an affordable supply of woody biomass generated through commercial activities. Officials of a rural school district told us that nearby lands are being cleared for development, and that a portion of the wood generated from land clearing is chipped by contractors for purchase by the school. Similarly, a Pennsylvania power plant uses wood from a combination of sources, including woody biomass from land-clearing operations that are, on average, more than 130 miles from the plant, according to a plant official [14]. This official told us that the developers clearing the land are required to dispose of the cleared material but are not allowed to burn or bury it, so they often are willing to partially subsidize removal and transportation costs in order to have an outlet for the material.

Forest management activities also contribute to the availability of an affordable supply of woody biomass. For example, small-diameter trees have been available to a large pulp and paper mill in Mississippi in part because of thinning activities by area landowners. In this area, as in much of the southeastern United States, forests are largely privately owned, and much of the forests are plantations meant for production. Small-diameter trees are periodically thinned from these forests to promote the growth of other trees, and traditionally have been sold for use in making pulp and paper. Officials at the Mississippi pulp and paper mill told us that these trees are a relatively inexpensive source of material compared with the cost of the material in other parts of the country because the structure of southeastern forests—with level terrain and extensive road access—reduces harvesting and hauling costs, in contrast to other parts of the country where steep terrain and limited road access may result in high harvesting and hauling costs.

Environmental Benefits and Other Factors Played a Role in the Use of Woody Biomass

Three users cited potential environmental benefits, such as improved forest health and air quality, as prompting their use of woody biomass; other users told us about additional factors that increased their use of woody biomass. Two users—the Montana school district and the coal-fired power plant in Colorado—started using woody biomass in part because of concerns about forest health and the need to reduce hazardous fuels in forest land; they also hoped that by providing a market for woody biomass, they could help stimulate thinning efforts. The Montana school district was the first of a series of Fuels for Schools projects intended to stimulate demand for woody biomass generated from forest fuels reduction, and the Colorado power plant began using woody biomass in an effort to contribute to the health of the forest by using material from nearby fuels reduction projects.

Air-quality concerns spurred the use of woody biomass at a Vermont power plant in our review. According to plant officials, the utilities that funded it were concerned about air quality and as a result chose to build a plant fired by wood instead of coal because wood emits lower amounts of pollutants. Other users cited the air-quality benefits of burning woody biomass under the controlled conditions of a boiler rather than burning it in the open air (whether through slash pile burning, prescribed burning, or wildland fire) because doing so generates significantly fewer emissions.

Finally, other factors and business arrangements specific to individual users encouraged the use of woody biomass, either by insulating users from the effects of changes in the price and availability of woody biomass or by enabling users to profitably add woody biomass use to their business. For example, an official at one wood-fired power plant told us that the plant has been able to operate because the plant's owners—a group of utilities—have the financial capacity, as well as a long-term outlook, to withstand short-term fluctuations in its profitability. Without this ownership, according to officials, the plant might have shut down during periods of decreased revenues resulting from variations in the price or availability of woody biomass. Another user, which chips wood for use as fuel in a nearby power plant, has an arrangement with the power plant under which the plant purchases the user's product at a price slightly higher than the cost the user incurred in obtaining and processing woody biomass, as long as the user's product is competitively priced and meets fuel-quality standards. The arrangement guarantees the user a long-term market for its product at a price that allows it to cover its costs. Three users whose operations include chipping woody biomass and other activities, such as commercial logging or sawmilling, told us that having these other operations within the same business is important because costs for equipment and personnel can be shared between the woody biomass chipping operation and the other activities.

Other users helped offset the cost of obtaining and using woody biomass by selling byproducts resulting from their use of the material. For example, one pulp and paper mill in our review sold turpentine and other byproducts that were produced during the production of pulp and paper, while another user—a wood-fired power plant—sold steam extracted from its turbine to a nearby food-canning factory. Other byproducts sold by users in our review included ash used as a fertilizer, bark for landscaping material, and sawdust used by particle board plants.

Challenges Faced by Woody Biomass Users Included Inadequate Supply and Costs Associated with Handling and Using the Material

Users in our review experienced factors that limited their use of woody biomass or made it more difficult or expensive to use, including insufficient supply and increased costs related to equipment and maintenance. Two users were unable to obtain a sufficient supply of woody biomass, and several more told us they had difficulty obtaining the material from federal lands. Several users also told us that, despite the economic advantages of using woody biomass in place of oil or natural gas, they had incurred costs that they would not have incurred had they burned oil or natural gas—including additional equipment for handling woody biomass and added operation and maintenance costs, such as costs arising from problems in storing and handling woody material.

Woody Biomass Was Not Always Sufficiently Available

Seven users in our review told us they had difficulty obtaining a sufficient supply of woody biomass, either because of constraints on the supply of the material or because of insufficient availability of loggers to collect it. Two users, both power plants, reported to us that they were operating at about 60 percent of their capacity because they were unable to obtain sufficient woody biomass or other fuel for their plants. Officials at both plants, each of which burned mostly woody biomass but also supplemented the material with mill residues and urban wood waste, told us that their shortages of wood were due at least in part to a shortage of nearby logging contractors. According to plant officials, the lack of logging contractors meant that nearby landowners were unable to carry out all of the projects they wished to undertake, resulting in what one plant official termed a "backlog of standing timber." While officials at one plant attributed the plant's shortage entirely to the insufficient availability of logging contractors, an official at the other plant stated that the lack of woody biomass from federal lands—particularly Forest Service lands—also was a significant problem. One plant reported taking a financial loss in each of the past 3 years, the result of operating below capacity.

The lack of supply from federal lands was a commonly expressed concern among the woody biomass users on the West Coast and in the Rocky Mountain region, with five of the seven users we reviewed in these regions (including one of the power plants running at about 60 percent capacity) telling us they had difficulty obtaining supply from federal lands. One such user ceased operations for several months because of an interruption in its supply of woody biomass from federal lands. Users with problems obtaining supply from federal lands generally expressed concern about the Forest Service's ability to conduct projects generating woody biomass; in fact, two users expressed skepticism that the large amounts of woody biomass expected to result from widespread thinning activities will ever materialize. One official stated, "We keep hearing about this coming 'wall of wood,' but we haven't seen any of it yet," adding that emphasizing uses for woody biomass without an adequate supply "is putting the cart before the horse" [15] Of the remaining six users in our review, one obtained about 5 percent of its woody biomass from federal lands while the other five used no federal woody biomass at the time of our review. In such cases, users obtained woody biomass from state or private lands, or relied on alternative wood materials such as sawmill residues or urban wood waste.

Users Choosing Woody Biomass over Oil or Natural Gas Made Additional Investments in Equipment and Incurred Additional Operations and Maintenance Costs

Several users in our review told us they incurred costs to purchase and install the equipment necessary to use woody biomass beyond the costs that would have been required for using fuel oil or natural gas. These costs included scales for weighing incoming material; truck tippers to assist in unloading material; wood-storage buildings or concrete pads for storing wood; chippers to chip the material to the proper size; and conveyors and other mechanisms for transporting the material to the boiler. Some users needed other equipment as well; an official at one location told us a front-end loader was dispatched every 45 minutes to push wood chips to a loading area, where a mechanical conveyor could pick the chips up. Figure 3 is a schematic of the equipment and process used by one user in our review.

The cost of this equipment varied considerably among users, in part as a result of the differences in the amount of wood consumed. For example, a school district burning about 850 tons of wood fuel per year reported spending about $385,000 for the necessary equipment, including the boiler, while a pulp and paper mill burning about 216,000 tons per year in its boiler—about 250 times the school district's annual consumption—reported investing $15 million in equipment necessary to use the material. Not all users reported additional substantial expenditures on equipment, however; one power plant burning wood mixed with coal told us that the only additional equipment it needed was a ramp for a front-end loader, which was constructed at minimal cost.

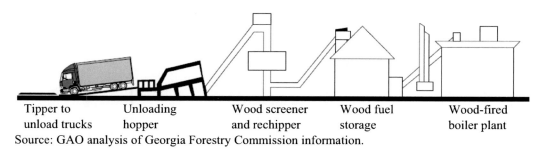

| Tipper to | Unloading | Wood screener | Wood fuel | Wood-fired |
| unload trucks | hopper | and rechipper | storage | boiler plant |

Source: GAO analysis of Georgia Forestry Commission information.

Figure 3. Equipment and Process for Using Wood Fuel at One Location.

Wood utilization also tended to increase operation and maintenance requirements for users. One power plant official told us that wood is more expensive to handle than coal, citing handling costs of $4.50 per ton for wood compared with $1.50 per ton for coal [16]. Wood also can create problems; for example, if wood chips are not properly sized, they can create blockages in machinery that require prompt action. During our visit to one facility, wood chips jammed on a conveyor belt, dumping wood chips over the side of the conveyor and requiring a maintenance crew member to manually clear the blockage. Figure 4 shows the crew member attempting to clear the blockage.

After one facility converted from natural gas and fuel oil to wood, it reported that the number of personnel needed to maintain its central heating plant nearly doubled, from about 8 or 9 to about 14 to 16. At another facility—the power plant mixing woody biomass with coal—an official told us that a wood blockage in the feed mechanism led to a fire in one of the plant's coal-storage units, requiring the plant to temporarily reduce its output of electricity and leading the plant to pay $9,000 to have its remaining stock of wood rechipped. Two users also reported spontaneous combustion in their wood storage piles that resulted from decaying wood.

Other issues specific to individual users also decreased woody biomass use or increased costs for using the material. For example, an official with one user, which chips small-diameter trees and sells the resulting chips to pulp and paper mills, told us that the pulp and paper mills prefer sawmill residues to chipped trees and will purchase his product only when sufficient sawmill residues are unavailable. This official told us that demand for his product has been so low in some years that he has operated his chip processor for only 6 months during the year. Another user, the Vermont wood-fired power plant, is required by the state to obtain 75 percent of its raw material by rail, in order to minimize truck traffic in a populated area.

According to plant officials, shipping the material by rail is more expensive than shipping by truck and creates fuel supply problems because the railroad serving the plant is unreliable and inefficient and experiences regular derailments. This same power plant is required by the state to restrict its purchases of woody biomass to material coming from forest projects that meet state-approved environmental standards. To ensure that it meets this requirement, the power plant employs four full-time foresters—an investment the plant would not have to make if it did not use woody biomass. Another power plant was required to obtain a new emissions permit in order to begin burning wood in its coal-fired system.

Figure 4. Maintenance Crew Member Clearing Wood Blockage in Conveyor Equipment.

An official at a third power plant told us that "woody biomass is expensive to harvest, process, transport, and handle—and it has only half the [energy] of coal"; [17] he summed up his concerns by stating, "Biomass energy is not the most efficient way to make electricity." However, he added that using woody biomass to make electricity provides benefits to society by consuming material that would otherwise be burned in the open or deposited in landfills.

Current Users' Experiences Offer Insights for Government Efforts to Expand the Use of Woody Biomass

Our findings offer several insights for promoting greater use of woody biomass. First, rather than helping to defray the costs of forest thinning, attempts to encourage the use of woody biomass may instead stimulate the use of other wood materials such as mill residues or commercial logging slash. Second, government activities may be more effective in stimulating woody biomass use if they take into account the extent to which a logging and milling infrastructure to collect and process forest materials is in place. And finally, the type of efforts employed to encourage woody biomass use may need to be tailored to the scale and nature of individual recipients' use.

It should be noted, however, that drawing long-term conclusions from the experiences of users in our review must be done with care because our review represents only a snapshot in time and a small number of woody biomass users. Changes in market conditions could have substantial effects on the options available to users and the materials they choose to consume, and the effects of changes in the market are complex and difficult to predict. For example, the price of fossil fuels such as natural gas plays a role in determining the cost-effectiveness of woody biomass use; if the price of natural gas were to rise, increased energy cost savings through woody biomass use might persuade more entities to convert to the material despite the up-front costs of conversion. On the other hand, if the cost of diesel fuel were to rise along with that of natural gas, the cost of harvesting and transporting woody biomass would increase because the machinery used to perform these tasks generally runs on diesel fuel—diminishing the advantages to be gained by using woody biomass.

Market Forces May Lead Wood Users to Forgo Small-Diameter Trees in Favor of Alternatives

One goal of the federal government's efforts to stimulate woody biomass use is to defray the cost to the government of thinning millions of acres of land at risk of wildland fire by creating a market for the resulting materials. Because a substantial component of these materials consists of small-diameter trees, it is important that government efforts include a focus on finding uses specifically for these trees. Without such a focus, efforts to stimulate woody biomass use may simply increase the use of mill residues or other wood materials—which several users told us were preferable to woody biomass for a variety of reasons—or slash from commercial logging operations.

Indeed, an indirect attempt to stimulate woody biomass use by one Montana user in our review led to the increased use of available mill residues instead. The Forest Service provided grant funds to finance the Montana school district's 2003 conversion to a wood heating system in order to stimulate the use of woody biomass in the area; the agency required as a condition of the grant that at least 50 percent of the district's fuel consist of woody biomass during the initial 2 years of the system's operation. Officials told us that the district complied with the requirement for those 2 years, but for the 2005-2006 school year, the district chose to use less expensive wood residues from a nearby log-home builder rather than woody biomass. The cost of these residues was $24 per ton, in contrast to the $36 per ton the district paid for woody biomass the previous year. A district official said that the district was willing to use woody biomass in the future if it could be obtained more cheaply than alternative materials [18]. The district was not alone among users in our review in its use of mill residues and other wood materials; eight users in our review used such materials in addition to, or instead of, woody biomass. Officials at one of these users—a pulp and paper mill—told us that they began their operation by using mill residues, switching to woody biomass only when competition for mill residues began driving up the price. Emphasizing users' preference for mill residues, a Forest Service official in Montana told us that his national forest sometimes has difficulty finding a market for woody biomass resulting from forest projects because the numerous log-home builders operating in the area offer a cheaper and more accessible source of wood in the form of mill residues.

This is not to say that the use of mill residues is entirely to the detriment of woody biomass. The use of mill residues can play an indirect role in facilitating woody biomass utilization by providing a market for the byproducts of industries using woody biomass directly, such as sawdust or other residues from small-log sawmills. The existence of a market for these byproducts can enhance the profitability of woody biomass users and, consequently, improve their ability to continue using woody biomass cost-effectively. In addition, the availability of both mill residues and woody biomass provides diversity of supply for users, allowing them to continue operations even if one source of supply is interrupted or becomes prohibitively expensive. Nevertheless, these indirect effects, even where present, may be insufficient to substantially influence the use of woody biomass.

Mill residues aside, even those users that consumed material we define as woody biomass, particularly those that used wood for fuel, often used the tops and limbs from trees harvested for merchantable timber or other uses rather than the small-diameter trees that contribute to the problem of overstocked forests. One woody biomass user in our review reported using only the slash from commercial logging rather than small-diameter trees, while another user reported that 80 percent of the woody biomass it used consisted of logging slash and 20 percent consisted of thinned small-diameter trees. Two users reported using residues from land-clearing operations conducted as part of commercial land development. Logging slash can be cheaper to obtain than small-diameter trees when it has been already removed from the forest by commercial logging projects; such projects often leave slash piles at roadside "landings," where trees are delimbed before being loaded onto log trucks. Unless woody biomass users specifically need small-diameter logs—for use in sawing lumber, for example—they may find it cheaper to collect slash piled in roadside areas than to enter the forest to cut and remove small-diameter trees. And while consuming logging slash may have environmental benefits—by, for example, decreasing smoke emissions by reducing the amount of slash burned in the open—it does not necessarily contribute to the government's goal of stimulating forest thinning or reducing thinning costs. Further, users' reliance on material whose cost of removal was subsidized by commercial activities suggests that, even if the government succeeds in stimulating a market for the woody biomass, it still may need to bear a substantial portion of thinning costs in order to make the material economically attractive for users.

The experience of the Montana school district also illustrates the unintended market consequences that may result from indirect attempts to stimulate woody biomass use. The school district is located in an area where several industries, including pulp and paper, plywood, and others, purchase commercially produced mill residues for their operations. By purchasing mill residues, the school district began competing for the same raw materials desired by these other industries. The impact on the market is likely to be small, as the school district uses only a small fraction of the wood used by these other industries. Nevertheless, in addition to spurring woody biomass use from forest-thinning operations, as originally envisioned by Forest Service officials, these grant funds also introduced more competition into an existing market for mill residues.

The Effectiveness of Efforts to Encourage Woody Biomass Use May Depend on the Presence of Other Wood-Related Industries

Government activities may be more effective in stimulating woody biomass use if they take into account the extent to which a logging and milling infrastructure is in place in potential users' locations. The availability of an affordable supply of woody biomass depends to a significant degree on the presence of a local logging and milling infrastructure to collect and process forest materials. Without a milling infrastructure, there may be little demand for forest materials, and without a logging infrastructure, there may be no way to obtain the materials. Indeed, officials at one power plant operating at less than full capacity because of a shortage of wood for the plant told us that the shortage was due to the lack of a local logging infrastructure—in other words, there simply weren't enough loggers to carry out the forest projects that nearby landowners wanted to undertake. The user said it was not cost-effective to obtain the material from more distant sources because of transportation costs.

Similarly, an official with the Nebraska state college in our review told us that the lack of a local logging infrastructure could potentially jeopardize the college's woody biomass use in the future. The college relied on logging slash from commercial loggers working nearby, but this official told us that the loggers were based in another state and the timber they were harvesting was hauled to sawmills over 100 miles away. The official said the loggers would prefer to work closer to the sawmills in order to reduce transportation costs, but could not find closer logging opportunities. According to the official, if more timber-harvesting projects were offered closer to the sawmills, these loggers would immediately move their operations—eliminating the nearby source of woody biomass available to the college.

In contrast, users located near a milling and logging infrastructure are likely to have more readily available sources of woody biomass. One Montana official told us that woody biomass in the form of logging slash is plentiful in the Missoula area, which is home to numerous milling and logging activities, and that about 90 percent of this slash is burned because it has no market. The presence of a logging and processing infrastructure, however, may increase the availability of mill residues, potentially complicating efforts to promote woody biomass use by offering more attractive alternative materials.

Efforts to Encourage Woody Biomass Use May Be More Effective If They Are Tailored to the Scale and Nature of Recipients' Use

Government activities may be more effective in stimulating woody biomass use if their efforts are tailored to the scale and nature of the users being targeted. Most of the large wood users we reviewed, such as pulp and paper mills or wood-fired power plants, were primarily concerned about supply, and thus might benefit most from federal efforts to provide a predictable and stable supply of woody biomass. Such stability might come, for example, from long-term contracts signed under stewardship contracting authority, which allows contracts of up to 10 years [19]. In fact, one company currently plans to build a $23 million woody biomass power plant in eastern Arizona, largely in response to the White Mountain stewardship project in the area, a thinning project expected to treat 50,000 to 250,000 acres over 10 years. Although the company is relying in part on $16 million in loan guarantees furnished by the Department of Agriculture, the assurance of supply offered by this long-term project was a key factor in the company's decision to build the power plant. Furthermore, a Department of Agriculture official told us that the assurance of supply also was critical to the department's decision to provide the loan guarantee. Similarly, in November 2005, officials of a South Carolina utility told us that the utility was planning to burn woody biomass resulting from thinning efforts in a nearby national forest, and was intending to purchase about 75,000 tons annually to burn along with coal in a coal-fired power plant. Although the utility did not yet have an agreement in place to purchase the woody biomass, the officials told us that the utility anticipated investing $4.4 million in wood-handling equipment and realizing substantial annual fuel and emissions cost savings [20]. The national forest expects to conduct several long-term thinning projects, and officials told us that the utility would not have considered making this investment in woody biomass use without this likelihood of a stable, long-term supply.

In contrast, small users we reviewed did not express concerns about the availability of supply, in part because their consumption was relatively small. However, three of these users relied on external financing for their up-front costs to convert to woody biomass use. Such users—particularly small, rural school districts or other public facilities that may face difficulties raising the capital to pay needed conversion costs—might benefit most from financial assistance such as grants or loan guarantees to fund their initial conversion efforts. And as we noted in our previous report on woody biomass, [21] several federal agencies provide grants for woody biomass use—particularly the Forest Service, which is, among other efforts, providing grants of between $50,000 and $250,000 to increase the utilization of woody biomass from or near national forest lands.

However, federal agencies must take care that their efforts to assist users are appropriately aligned with the agencies' own interests, and that their efforts do not create unintended consequences. For example, while individual grant recipients might reap substantial benefits from their ability to use woody biomass—through fuel cost savings, for example, as demonstrated by several users in our review—benefits to the government, such as reduced thinning costs, are uncertain. Without such benefits, agency grants may simply increase agency outlays but not produce comparable savings in thinning costs. The agencies also risk adverse ecological consequences if their efforts to develop markets for woody biomass result in these markets inappropriately influencing land management decisions. As noted in our prior report on woody biomass, agency and nonagency officials cautioned that efforts to supply woody biomass in response to market demand rather than ecological necessity might result in inappropriate or excessive thinning.

CONCLUDING OBSERVATIONS

The variety of factors influencing woody biomass use among users in our review—including regulatory, geographic, market-based, and other factors—suggests that the federal government may be able to take many different approaches as it seeks to stimulate additional use of the material. However, because these approaches have different costs, and likely will provide different returns in terms of defraying thinning expenses, it will be important to identify what kinds of mechanisms and what types of resource investments are most cost-effective in different circumstances. This will be a difficult task, given the variation in different users' needs and available resources, differences in regional markets and forest types, and the multitude of available alternatives to woody biomass. Nevertheless, if federal agencies are to maximize the long-term impact of the millions of dollars being spent to stimulate woody biomass use, they will need to design approaches that take these elements into account rather than using boilerplate solutions.

APPENDIX I
CHARACTERISTICS OF WOODY BIOMASS
USERS INCLUDED IN OUR REVIEW

Table 1 provides information on the type and amount of wood fuel consumption reported by each woody biomass user in our review. This information is based on the amount of wood used in the last full year for which complete data were available.

Table 1. Characteristics of Woody Biomass Users Included in Our Review

Woody biomass user	Primary use	Wood used per year (bone dry tons)[b]	Woody biomass as a percentage of all wood used	Logging slash as a percentage of woody biomass used[c]	Percent of woody biomass obtained from federal lands
Contractor, MI	Wood fuel	110,000	100	93	5
Contractor, MT	Wood fuel	75,000	67	80	10
Contractor, OR	Chips for pulp	60,000	100	Not provided	45
Power plant, CA	Electricity generation	126,000	67	Not provided	49
Power plant, CO	Electricity generation	760[d]	100	0	100
Power plant, PA	Electricity generation	140,000	74	100	0
Power plant, VT	Electricity generation	180,000	75	100	0
Pulp and paper mill, MS	Pulp and paper; process steam	1,600,000	50	Not provided	0
Pulp and paper mill, MT	Pulp and paper; process steam	966,000	30	30	less than 5
Rural school district, MT	Building heat	490	67	40	100
Rural school district, PA	Building heat	850	100	100	0
State college, NE	Building heat	6,000	100	100	0
State hospital, GA	Building heat	9,000	0[e]	N/A	N/A
Total		3,274,100			

Source: GAO analysis of users' data.

[a]Figures were derived from information provided by users in our review.

[b]One bone-dry ton represents 1 ton of wood at 0 percent moisture content; green wood generally contains about 50 percent moisture. This column represents all types of wood used, including alternative materials such as mill residues, during the most recent year for which complete information was available.

[c]Figures presented in this column include residues generated by commercial land clearing activities as well as logging slash generated by commercial logging operations.

[d]This power plant mixed woody biomass with coal on a trial basis to determine its feasibility. The amount of woody biomass the plant burned represents a small fraction of the plant's annual consumption of coal.

[e]The state hospital in Georgia has historically used woody biomass, but had not done so during the most recent year for which complete information was available.

APPENDIX II
OBJECTIVES, SCOPE AND METHODOLOGY

The objectives of our review were to (1) identify key factors facilitating the use of woody biomass among selected users, (2) identify challenges these users have faced in using woody biomass, and (3) discuss the insights our findings offer for promoting greater use of woody biomass. To meet these objectives, we reviewed the operations of 13 public and private organizations throughout the United States that use woody biomass to make a variety of products.

Because no comprehensive list of woody biomass users exists, we asked knowledgeable federal and nonfederal officials to identify woody biomass users. As part of these interviews, we asked for names of additional officials—regardless of location or agency affiliation—who could provide additional information about, or insights into, woody biomass users. Federal officials we met with included various officials from the Department of Agriculture's Forest Service, Department of Energy, and Department of the Interior. We also contacted nonfederal officials, including representatives of the Appalachian Hardwood Center, Biomass Energy Resource Center, Bitter Root Resource Conservation and Development Council, Center for Biological Diversity, Montana Community Development Corporation, National Association of Conservation Districts, Natural Resources Defense Council, Society of American Foresters, Southern Alliance for the Utilization of Biomass Resources, USA Biomass Power Producers Alliance, and Wilderness Society. We asked these federal and nonfederal officials to identify woody biomass users across the United States. We continued this expert referral technique until the references we received for woody biomass users became repetitive. We also reviewed documents that identified possible woody biomass users and provided background information about woody biomass use.

From the several hundred entities that were reported to us as using woody biomass, we selected for further review a nonprobability sample of 14 woody biomass users from different industries and geographic locations [22].These users produced a range of different products from woody biomass, such as building heat, electricity, pulp, paper, and wood fuel, and were located in various geographic locations around the country. Of these users, 13 participated in our review; the remaining user, a sawmill using small-diameter logs to make lumber, did not respond to our request to participate. The woody biomass users we reviewed included:

- a state college in Nebraska,
- a state hospital in Georgia,
- two rural school districts in Montana and Pennsylvania,
- two pulp and paper mills in Mississippi and Montana,
- three logging and wood products operations in Michigan, Montana, and Oregon, and
- four electric power producers in California, Colorado, Vermont, and Pennsylvania.

The general locations of the users we reviewed are shown in figure 5.

We then developed a structured interview guide to review the operations of the 13 woody biomass users and to obtain general information about their operations. Because the practical difficulties of developing and administering a structured interview guide may introduce errors—resulting from how a particular question is interpreted, for example, or from

differences in the sources of information available to respondents in answering a question—
we included steps in the development and administration of the guide for the purpose of
minimizing such errors.

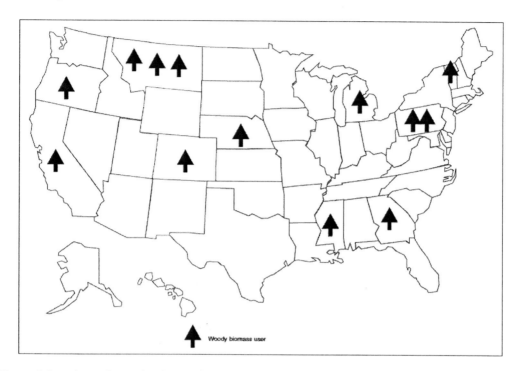

Figure 5. Locations of Woody Biomass Users We Reviewed.

We pretested the guide at one location and conducted a second pretest by telephone. We
also provided a draft version of the guide to federal officials knowledgeable about woody
biomass in order to obtain their comments on the draft. Based on these steps, we modified the
structured interview guide to reflect questions and comments we received.

Factors Facilitating the Use of Woody Biomass among Selected Users

To collect information about the factors that facilitate woody biomass use, we used our
structured interview guide to obtain information about the 13 users in our review, including
the types and amount of woody biomass used, when their woody biomass use began, and the
type of materials that woody biomass use replaced. We also asked users about economic
factors that facilitated their use of woody biomass, such as the cost and availability of their
supply. During discussions, we also gathered users' opinions about factors that might increase
their use of woody biomass. To corroborate the information we gathered through interviews,
we compared interviewees' responses with other information we reviewed, when available,
such as contracts, third-party evaluations of user activities, financial analyses, and the like.
Because the documentary evidence we reviewed generally agreed with the information
provided by woody biomass users, we believe the data are sufficiently reliable to be used in

providing descriptive information on the factors facilitating woody biomass use by users in our review.

Challenges Faced by Selected Users of Woody Biomass

We also used our structured interview guide to ask the 13 users about challenges they faced or other factors that might diminish their use of woody biomass. For example, we asked users about the affordability of their supply of woody biomass and the farthest distance from which they can affordably obtain it, and gathered users' opinions about factors that might diminish their use of woody biomass. To corroborate the information we gathered through interviews, we compared interviewees' responses with other information we obtained, when available—again including documentation such as contracts, third-party evaluations of user activities, financial analyses, and the like. Because the documentary evidence we reviewed generally agreed with the information provided by woody biomass users, we believe the data are sufficiently reliable to be used in providing descriptive information on challenges associated with the use of woody biomass by users in our review.

Insights Offered by Our Findings

To describe the insights offered by our findings, we relied principally on the information gathered during our discussions with woody biomass users. In addition, we used information gathered from interviewing potential and current users as well as agency officials and others knowledgeable about woody biomass use, including information gathered during our prior review of woody biomass. Our intent was to highlight issues that we observed in our review of current woody biomass users and that we believe should be considered by those seeking to develop a market for woody biomass.

We performed our work from May 2005 through January 2006 in accordance with generally accepted government auditing standards.

GAO COMMENTS

(1) We agree that the multiple benefits listed by the department may indeed flow from the increased use of woody biomass. However, the objectives of our review were to evaluate the experiences of individual users, not to identify the general benefits of using woody biomass. In instances in which users cited certain benefits facilitating or resulting from their use of woody biomass, we included such benefits in our report.

(2) We have modified our draft to include reference to the department's contracting clause.

(3) We understand that projects focusing on after-treatment stand conditions may generate a variety of materials, including commercial sawtimber and lower-value materials, and that the commercial component of these materials can help offset project costs. However, while it is not possible to separate the two issues entirely, the

focus of our report—embodied in our definition of woody biomass—is on small-diameter trees and other traditionally noncommercial material.

(4) The focus of our report goes beyond heat and electricity production, and includes two users manufacturing pulp and paper from woody biomass and three contractors processing woody biomass for other users. Nevertheless, we recognize that users in other industries—or even other users in the same industries we examined—may offer additional insights into expanding the use of woody biomass.

(5) Large-scale operations or widespread penetration of small wood industries might, as the department suggests, create competition for the materials and increase the value of small-diameter trees. However, while such a scenario may come about in the long term, our intent was to provide insights and information applicable to the current situation. Similarly, the scope of our report generally is limited to the experiences of individual users in our review, not to broader efforts such as the northern California woody biomass thinning program.

(6) We do not necessarily look at mill residues as a negative influence on using woody biomass—in fact, we acknowledge that the use of mill residues and other sources of wood can benefit woody biomass utilization in several ways. However, we do believe that it can serve as a complicating factor in the government's efforts to stimulate the use of small-diameter trees specifically.

(7) We have modified our draft to reflect the current expiration date of 2013 for Forest Service and Bureau of Land Management stewardship contracting authority, as well as to note the related authorities available to other Interior agencies.

(8) While the department's suggestions regarding stewardship contracting and market and tax incentives are beyond the scope of our review, the agencies or the Congress may wish to consider these options as they evaluate the success of contracting and grant programs.

REFERENCES

[1] Although biomass can be considered any sort of organic material—including trees, grasses, agricultural crops, and animal wastes—the term woody biomass in this report refers to small-diameter trees and other traditionally noncommercial material cut as part of thinning, harvesting, or other activities on forests or rangelands. For the purposes of this report, we distinguish woody biomass from other wood residues such as sawmill residues or urban wood waste.

[2] See GAO, *Natural Resources: Federal Agencies Are Engaged in Various Efforts to Promote the Utilization of Woody Biomass, but Significant Obstacles to Its Use Remain*, GAO-05-373 (Washington, D.C.: May 13, 2005).

[3] Pub. L. No. 106-224, Title III, 114 Stat. 428 (2000), as amended.

[4] Pub. L. No. 108-148, Title II, 117 Stat. 1901 (2003).

[5] Pub. L. No. 108-447, 118 Stat. 3076 (2004).

[6] Pub. L. No. 109-58, § 210, 119 Stat. 658 (2005).

[7] Resource Conservation and Development Councils are part of the Resource Conservation and Development Program, managed by the Department of Agriculture's

Natural Resources Conservation Service. The program is intended to encourage and improve the capability of state and local units of government and local nonprofit organizations in rural areas to plan, develop, and carry out programs for resource conservation and development.

[8]   Dollars are unadjusted for inflation.

[9]   Another user in our review that chips woody biomass into raw material for pulp and paper plants has obtained a federal grant to build a sawmill capable of processing small-diameter logs into lumber. However, this sawmill was not yet in operation at the time of our review.

[10]   A megawatt is a unit of power equal to 1 million watts, or enough electricity to power about 750 homes at any given time.

[11]   Pub. L. No. 95-617, 92 Stat. 3117 (1978).

[12]   States set rates, pursuant to general regulations issued by the Federal Energy Regulatory Commission, based on the buyer's "avoided cost." Avoided costs are the energy and facilities costs that would have been incurred by the purchasing utility if that utility had to provide its own generating capacity. According to the commission, while it provides general avoided cost regulations, states set rates that often are above market rates.

[13]   The 730 megawatt hours represent a small fraction of the plant's output, nearly all of which is generated by burning coal. However, the plant determined—and a third-party certifying body agreed—that 730 megawatt hours could be directly attributed to burning woody biomass.

[14]   Other sources of material used by the plant, according to this official, include slash from conventional logging, chips from sustainably managed forestry operations, sawmill waste, and urban wood waste.

[15]   In commenting on a draft of this report, officials from both the Department of the Interior and the Forest Service stated that their agencies are seeking to increase the availability of woody biomass from federal lands. Interior officials commented that the department has established a contract clause, to insert in all solicitations and contracts through which biomass is expected to be generated, allowing the use of all material so long as it is ecologically appropriate and in accordance with the law. Interior stated that this clause could increase the availability and affordability of woody biomass from Interior lands. Forest Service officials told us that their agency is seeking new opportunities for providing a reliable and consistent supply of woody biomass, including working with Interior to streamline processes for developing stewardship contracts and agreements. Stewardship contracting involves the use of any of several contracting authorities on the part of the Forest Service and Interior's Bureau of Land Management, including the ability to exchange goods for services and to enter into contracts of up to 10 years. Stewardship contracting authority expires in 2013. Other Interior agencies, including the Fish and Wildlife Service, have related authorities; see, e.g., 50 C.F.R. § 29.5.

[16]   It should be noted that this plant used substantially more coal than wood, and, as a result, the lower handling cost for coal might be attributable in part to economies of scale.

[17]   According to the Forest Service's Forest Products Laboratory, the typical heating value of wood ranges from about one-quarter to one-half that of bituminous coal, depending

on the moisture content of the wood. However, some power plants use lower grades of coal that can have heating values comparable to that of oven-dried wood.

[18] Subsequent to our review of the school district's operations, the district obtained about 550 tons of woody biomass (about 75 percent of its annual consumption) from a nearby thinning project at a price of $10 per ton. This price represents only a fraction of the material's processing and handling costs, most of which were borne by the Forest Service.

[19] For a description of agency use of stewardship contracting authority, see GAO, *Federal Land Management: Additional Guidance on Community Involvement Could Enhance Effectiveness of Stewardship Contracting*, GAO-04-652 (Washington, D.C.: June 14, 2004).

[20] Utility officials told us they estimate saving about $1.4 million annually, but noted that this figure could vary significantly depending on prices for coal and wood and on the amount of wood used.

[21] GAO-05-373.

[22] Results from nonprobability samples cannot be used to make inferences about a population, because in a nonprobability sample, some elements of the population being studied have no chance or an unknown chance of being selected as part of the sample.

In: Environmental Research Advances   ISBN: 978-1-60021-762-3
Editor: Peter A. Clarkson, pp. 45-75   © 2007 Nova Science Publishers, Inc.

Chapter 2

THE NATIONAL FOREST SYSTEM ROADLESS AREAS INITIATIVE[*]

Pamela Baldwin and Ross Gorte

ABSTRACT

The roadless areas of the National Forest System have received special management attention for decades. In part to recognize the importance of national forest roadless areas for many purposes and in part because making project decisions involving roadless areas on a forest-by-forest basis was resulting in controversy and litigation that consumed considerable time and money, the Clinton Administration established a new nationwide approach to the management of the roadless areas in the National Forest System. A record of decision (ROD) and a final rule were published on January 12, 2001, that prohibited most road construction and reconstruction in 58.5 million acres of inventoried forest roadless areas, with significant exceptions. Most timber cutting in roadless areas also was prohibited, with some exceptions, including improving habitat for threatened, endangered, proposed, or sensitive species, or reducing the risk of wildfire and disease. With some exceptions, the new prohibitions would have applied immediately to the Tongass National Forest in Alaska.

The Bush Administration initially postponed the effective date of the roadless area rule, then decided to allow it to be implemented while proposing amendments. However, the Federal District Court for Idaho preliminarily enjoined its implementation. The Ninth Circuit reversed the Idaho district court, but on July 14, 2003, the Federal District Court for Wyoming again permanently enjoined implementation of the Rule. This holding was appealed but dismissed as moot by the 10th Circuit. A final rule "temporarily" exempting the Tongass National Forest (Alaska) from the roadless rule was published December 30, 2003. On July 16, 2004, a new general roadless rule was proposed and a new final rule was published on May 13, 2005. The new rule eliminates the need for a separate Tongass rule, and replaces the Clinton roadless rule with a procedure whereby the governor of a state may petition the Secretary of Agriculture to promulgate a special rule for the management of roadless areas in that state, and make recommendations for that management. The new rule contains no standards by which the Secretary is to review a state's recommendations, and does not address the weight to be given to the views of

---

[*] Excerpted from CRS Report RL30647, dated September 7, 2006.

local residents versus those of out-of-state residents with an interest in the public lands. Five states, including New Mexico and California, have filed petitions so far and those of Virginia, North Carolina, and South Carolina have been approved. Other states have until November 13, 2006, to file petitions under the special procedure or may seek a rule change thereafter under 7 C.F.R. § 1.28.

A new final forest planning rule was also published on January 5, 2005. The rule does not address roadless areas at all, but does require review of areas that might be suitable for wilderness designation when plans are revised.

This article traces the development and content of the roadless area rules and directives, describes the statutory background, reviews recent events, and analyzes some of the legal issues.

THE ROADLESS AREAS AND RELATED INITIATIVES

Background

The Clinton Administration undertook a series of actions affecting the roadless areas of the National Forest System (NFS).[1]. More particularly, new rules were finalized with respect to: (1) the roadless areas as such; (2) the NFS roads that make up the Forest Development Transportation System, and (3) the planning process of the Forest Service (FS). The provisions of these three new rules were intertwined and each part affected the others. The new roadless area rules were issued in light of the importance of the roadless areas for many forest management purposes and to the American public, and because addressing projects in roadless areas on a forest-by-forest basis as part of the usual planning process was resulting in controversy, conflict, and the expenditure of time and money on appeals and litigation, such that national-level guidance regarding projects in roadless areas was deemed advisable.

The Clinton Administration roadless area proposals were praised by some, criticized by some for not being far-reaching enough, and criticized by others as being too restrictive, creating "de facto wilderness," and being procedurally flawed. Several lawsuits were filed challenging the adequacy of the information provided the public, the opportunity to comment, and other aspects of the rulemaking. The Bush Administration initially postponed the effective date of the roadless area rule, but then decided to implement it while considering changes. Implementation of the rule was enjoined on May 10, 2001, but this district court decision was reversed and remanded by the 9th Circuit. However, the rule was again enjoined by the Wyoming District Court on July 14, 2003, on NEPA grounds and because the court concluded it created de facto wilderness. This decision was appealed by intervenors to the 10th Circuit but was dismissed as moot in light of the fact that the challenged roadless rule has now been replaced.

The Bush Administration put into effect an initial set of interim directives regarding roadless area management, solicited comments retroactively on the interim directives and on the management of roadless areas in general, and finalized a directive altering the requirements for preparation of NEPA documents in some instances. A rule exempting the Tongass National Forest in Alaska from the Clinton rule was then published, pending other possible rule changes to address the application of the roadless rule in Alaska. On July 16, 2004, new roadless rules were proposed that would eliminate the previous rule and provide a new procedure for the governor of a state to petition the Secretary of Agriculture for a state-wide rule on the management of roadless areas. In the meantime, roadless areas would be managed under the interim direction, with some modifications. New final forest planning rules were published on January 5, 2005, and a new round of interim directives was published. A new final roadless area rule was published on May 13, 2005 that returns roadless area management to the general forest planning processes, unless the governor of a state petitions for a special state-specific rule [2]. Five states — Virginia, North Carolina, South Carolina, New Mexico, and California — have now petitioned for special state-wide rules, and the first three of these have been approved. Other states have until November 13, 2006, to file petitions under the roadless rule procedures or may petition for a rule change under 7 C.F.R. § 1.28 thereafter.

This article focuses on the roadless areas initiative, describes the statutory background, summarizes and provides citations for the various rules and subsequent actions, and analyzes some of the legal issues in connection with the roadless areas.

Roadless Areas

On October 13, 1999, President Clinton directed the Secretary of Agriculture, acting through the Forest Service, to develop regulations to provide "appropriate long-term protection for most or all of the currently inventoried 'roadless' areas, and to determine whether such protection is warranted for any smaller roadless areas not yet inventoried." [3] A Notice of Intent to complete an environmental impact statement (EIS) on alternatives for protection of NFS roadless areas was published on October 19, 1999; [4] a draft EIS (DEIS) was issued in May, 2000, and accompanying regulations were proposed on May 10, 2000;[5] and a final environmental impact statement (FEIS) was issued on November 13, 2000. A record of decision (ROD) and final rules were issued on January 12, 2001, to be effective on March 13, 2001 [6]. The rules were issued in light of the importance of the roadless areas for various forest management purposes and to the American public, and because addressing projects in roadless areas on a forest-by-forest basis as part of the planning process was resulting in controversy, conflict, and the expenditure of a great deal of time and expense on appeals and litigation, such that national-level guidance was deemed advisable [7]

The ROD and final rule would have (1) prohibited, with significant exceptions, new roads in inventoried roadless areas; (2) prohibited most timber harvests in the roadless areas, but allow cutting under specified circumstances; and (3) applied the same prohibitions to the Tongass National Forest in Alaska, but allowed certain road and harvest activities already in the pipeline to go forward. The details of the final rule will be discussed below.

Roads

In related actions, the Forest Service on January 28, 1998, issued an Advance Notice of Proposed Rulemaking to revise its Forest Development Transportation System regulations related to roads in the NFS, [8] and also proposed an interim rule to temporarily suspend road construction and reconstruction in certain NFS unroaded areas [9]. On February 12, 1999, the agency published a final interim rule that temporarily suspended road construction and reconstruction in unroaded areas, and provided certain procedures related to such areas [10]. A proposed rule [11] and proposed administrative policy [12] regarding the Forest Development Transportation System were published on March 3, 2000. Final Roads rules (36 C.F.R. §212) and a transportation policy were published on January 12, 2001, both effective on that date.[13] (Note that the final roadless area management rule also was published on that date.) Certain terms were changed in the final rule, [14] and the policy provided new direction to be contained in the Forest Service Manual that emphasizes the maintenance and decommissioning of existing roads rather than the construction of new roads. The policy addressed when and how to conduct roads analyses, required that a compelling need for a new road be demonstrated, and also required an economic analysis that addressed both initial and long-term costs, a scientific analysis, and a full EIS before a road could be built in roadless areas. The new final policy was to supersede the interim policy except with respect to roads in the Tongass National Forest, in which forest the interim policy would continue to govern the activities that are permitted to go forward. These policies and interim direction have now been changed under the Bush Administration.

Under 36 C.F.R. § 212.5(b), [15] the focus is on providing and maintaining the minimum forest transportation system needed for safe and efficient travel and for the administration, utilization, and protection of NFS lands. This is to be determined by science-based roads analysis at the appropriate scale and is to minimize adverse environmental impacts. Unneeded roads would be decommissioned and the roadbeds restored. The economic and ecological effects of roads would be analyzed as part of an interdisciplinary, "science-based" process in which the public would be engaged. Until the new road inventories and analyses are completed, interim requirements would pertain and a compelling need for new roads would have to be demonstrated. These rules are still in effect, though new rules have been proposed.

Planning

On a third track, the Forest Service on November 9, 2000, issued final new planning regulations, effective on that date [16]. However, new planning regulations were again proposed on December 6, 2002, and the date for compliance with the 2000 planning regulations was extended, [17] — which had the effect of restoring the 1982 planning regulations until new planning regulations were finalized. New regulations were published on January 5, 2005 [18].

ROADLESS AREAS: STATUTORY BACKGROUND

In considering the roadless area initiatives, a review of the most relevant portions of the statutes that govern the management of the NFS may be helpful.

The principal forest management statutes relevant to analysis of the Roadless Area Initiative are the Organic Act of 1897, [19] the Multiple-Use Sustained-Yield Act of 1960, [20] and the National Forest Management Act of 1976. [21]. The 1897 Act directs that the national forests be managed to improve and protect the forests or "for the purpose of securing favorable conditions of water flows, and to furnish a continuous supply of timber for the use and necessities of citizens of the United States ...." [22]. The 1897 Act also authorizes the Secretary to issue regulations to "regulate the occupancy and use of the forests and to preserve them from destruction ...." [23]

Over the years, many uses of the national forests in addition to timber and watershed management have been allowed administratively. Statutorily, the Multiple-Use Sustained-Yield Act of 1960 (MUSYA) expressly recognizes and authorizes the "multiple use" of the forests, a term MUSYA defines as the management of all the various renewable surface resources of the national forests "in the combination that will best meet the needs of the American people" and recognizes that "some land will be used for less than all of the resources ... without impairment of the productivity of the land, with consideration being given to the relative values of the various resources, and not necessarily the combination of uses that will give the greatest dollar return or the greatest unit output." [24]. MUSYA states that the national forests are established and shall be administered for their original purposes and also for "outdoor recreation, range, timber, watershed, and wildlife and fish purposes"[25] and that "[t]he establishment and maintenance of areas of wilderness are consistent with the purposes and provisions of [the act.]"[26]. This latter language, which preceded enactment of the 1964 Wilderness Act , [27] recognized that the FS had been managing some forest areas as administrative wilderness or natural areas. What constitutes the most desirable combination of uses for a forest has been hotly debated for decades.

MUSYA also requires "sustained yield," which is defined as the "achievement and maintenance in perpetuity of a high-level annual or regular periodic output of the various renewable resources of the national forests without impairment of the productivity of the land." [28]. How much is a "high-level annual or regular periodic output" of forest resources that does not impair the productivity of the land has also been the subject of much debate.

The National Forest Management Act of 1976 (NFMA) set out additional provisions on the management of the national forests that include direction for developing land and resource management plans. NFMA directs that regulations be adopted to guide forest planning and accomplish specific goals set by the Congress, including insuring consideration of the economic and environmental aspects of various systems of renewable resource management including "silviculture and protection of forest resources; to provide for outdoor recreation (including wilderness), range, timber, watershed, wildlife, and fish; and providing for diversity of plant and animal communities"[29].

The roadless areas in the National Forest System have long received special management attention. Beginning in 1924, long before the enactment of MUSYA, the FS managed many forest areas as natural, primitive, or wilderness areas — a practice expressly approved in MUSYA. More permanent, congressionally approved statutory wilderness areas were provided for in the 1964 Wilderness Act, [30] which established the National Wilderness Preservation System. The Wilderness Act directed review of FS-designated primitive areas and other larger roadless areas to consider their suitability for inclusion in the national wilderness system. This review was carried out and expanded (with respect to the national forests) in the Roadless Area Review and Evaluation or "RARE" studies, which expanded on studies begun before enactment of the 1964 Wilderness Act. Roadless areas inventoried either as part of the RARE studies or as part of subsequent reviews during the NFMA planning process are the "inventoried" roadless areas referred to in the October 19, 1999 Notice. Congress has designated many additional wilderness areas since 1964, but, under the statutes summarized above, especially the MUSYA, the FS may still manage parts of the national forests as natural, primitive, or wildlife areas, which might be characterized as "administrative wilderness" areas.

The management of the roadless areas of the NFS is of great interest to both wilderness proponents and to opponents of additional natural or wilderness area protection. Proponents of additional protection point to the many purposes the roadless areas serve, including water quality protection, backcountry recreation, and habitat for wildlife; opponents assert that the formal congressional wilderness review and designation process sets aside adequate natural areas and the remaining areas should be available for timber harvesting, mining, developed recreation, and other uses.

The FS identified approximately 58.5 million acres of inventoried roadless areas, roughly one-third of all NFS lands. Road building is not allowed in 20.5 million acres of this total under current plans. Roads are also currently prohibited in an additional 42.4 million acres of Congressionally-designated areas such as Wilderness or Wild and Scenic River corridors. There are approximately 386,000 miles of FS and other roads in the NFS. The explanatory material in the final rulemaking states that roadless areas provide significant opportunities for dispersed recreation, are sources of public drinking water, and are large undisturbed landscapes that provide open space and natural settings, serve as a barrier against invasive plant and animal species, are important habitat, support the diversity of native species, and provide opportunities for monitoring and research [31]. In contrast, the explanatory material continues, installing roads can increase erosion and sediment yields, disrupt normal water flow processes, increase the likelihood of landslides and slope failure, fragment ecosystems, introduce non-native species, compromise habitat, and increase air pollution [32].

THE FINAL CLINTON ADMINISTRATION ROADLESS AREA RULE

The final roadless area rule put in place by the Clinton Administration was more restrictive in several respects than was either the proposed roadless rule or the preferred alternative set out in the FEIS. With some exceptions, the final rule imposed immediately-effective, national-level, Service-wide, limitations on new road construction and reconstruction in the inventoried roadless areas throughout the NFS, and also imposed nationwide prohibitions on timber harvesting in those areas, with some exceptions. The regulations were to apply immediately to the Tongass National Forest in Alaska, although certain activities already in the planning stages in that Forest were allowed to go forward.

The final rule prohibited new road construction and reconstruction, but with some significant exceptions. The exceptions were if:

(1) A road is needed to protect public health and safety in cases of an imminent threat of flood, fire, or other catastrophic event that, without intervention, would cause the loss of life or property;

(2) A road is needed to conduct a response action under the Comprehensive Environmental Response, Compensation, and Liability Act (CERCLA) or to conduct a natural resource restoration action under CERCLA, Section 311 of the Clean Water Act, or the Oil Pollution Act;

(3) A road is needed pursuant to reserved or outstanding rights, or as provided for by statute or treaty;

(4) Road realignment is needed to prevent irreparable resource damage that arises from the design, location, use, or deterioration of a classified road and that cannot be mitigated by road maintenance. Road realignment may occur under this paragraph only if the road is deemed essential for public or private access, natural resource management, or public health and safety;

(5) Road reconstruction is needed to implement a road safety improvement project on a classified road determined to be hazardous on the basis of accident experience or accident potential on that road;

(6) The Secretary of Agriculture determines that a Federal Aid Highway project, authorized pursuant to Title 23 of the United States Code, is in the public interest or is consistent with the purposes for which the land was reserved or acquired and no other reasonable and prudent alternative exists; or

(7) A road is needed in conjunction with the continuation, extension, or renewal of a mineral lease on lands that are under lease by the Secretary of the Interior as of January 12, 2001 or for a new lease issued immediately upon expiration of an existing lease. Such road construction or reconstruction must be conducted in a manner that minimizes effects on surface resources, prevents unnecessary or unreasonable surface disturbance, and complies with all applicable lease requirements, land and resource management plan direction, regulations, and laws. Roads constructed or reconstructed pursuant to this paragraph must be obliterated when no longer needed for the purposes of the lease or upon termination or expiration of the lease, whichever is sooner.

Maintenance of classified roads was permissible in inventoried roadless areas.

The cutting, sale, or removal of timber from inventoried roadless areas also was prohibited unless one of specified circumstances exists, and the expectation was expressed that cutting would be infrequent. The proposed regulations had allowed timber to be cut for "stewardship" purposes, but the final regulation eliminated the use of that ambiguous term in favor of specifying the purposes for which cutting could be allowed. Cutting of small diameter trees was permissible if doing so would maintain or improve one or more of the roadless area characteristics and would:

- improve habitat for species that are listed as threatened or endangered under the Endangered Species Act or are proposed for listing under that Act, or which are sensitive species; or
- maintain or restore ecosystem composition and structure, such as to reduce the risk of uncharacteristic wildfire effects [33]

Other cutting could be permitted if incidental to the implementation of a management activity that was not otherwise prohibited; if needed and appropriate for personal or administrative use in accordance with 36 C.F.R. § 223 (the regulations on sale and disposal of timber); or if roadless characteristics had been substantially altered in a portion of an inventoried roadless area due to the construction of a classified road and subsequent timber harvest before January 12, 2001. In this last instance, timber could only be cut in the substantially altered portion of the roadless area [34]

The new roadless area rule expressly would not have revoked, suspended, or modified any permit, contract, or other legal instrument authorizing the occupancy and use of NFS lands that was issued before January 12, 2001, nor would it have revoked, suspended, or modified any project or activity decision made prior to January 12, 2001 [35]. The rule would not have applied to roads or harvest in the Tongass National Forest if a notice of availability of a draft environmental impact statement for the activities had been published in the Federal Register before January 12, 2001 [36]. These provisions would have grandfathered the activities addressed, but otherwise the new rule would have applied to the Tongass immediately [37].

The explanatory material accompanying the Clinton Administration's planning rule of November 9, 2000, indicated that it was very similar to the proposed roadless area rule and also stated that the "final planning rule clarifies that analyses and decisions regarding inventoried roadless areas and other unroaded areas, other than the national prohibitions that may be established in the final Roadless Area Conservation Rule, will be made through the planning process articulated in this final rule. Under this final rule, the responsible official is required to evaluate inventoried roadless areas and unroaded areas and identify areas that warrant additional protection and the level of protection to be afforded" [38].

Therefore, possible *additional* restrictions on use of the roadless areas beyond those provided by the national rule would be developed as part of the planning process. The materials also compared particular parts of the proposed roadless areas rule with the final planning rule. It appears that the final planning regulations are less specific with respect to roadless area reviews than were the proposed regulations. As noted, the final rule eliminated the separate treatment of roadless area reviews within that rule.

CAN "DE FACTO" WILDERNESS AREAS BE CREATED ADMINISTRATIVELY?

Some have asserted that the management changes involved in the roadless area initiative would amount to "de facto" wilderness, and that only Congress can designate wilderness areas.

The explanatory material with the final Clinton roadless area regulation stated that the regulation preserves "multiple use" management and that a wide range of multiple uses were permitted in inventoried roadless areas subject to the management direction in forest plans before the roadless rule, and that a wide range of multiple uses would still be allowable under the new rule.

Under this final rule, management actions that do not require the construction of new roads will still be allowed, including activities such as timber harvesting for clearly defined, limited purposes, development of valid claims of locatable minerals, grazing of livestock, and off-highway vehicle use where specifically permitted. Existing classified roads in inventoried roadless areas may be maintained and used for these and other activities as well. Forest health treatments for the purposes of improving threatened, endangered, proposed, or sensitive species habitat or maintaining or restoring the characteristics of ecosystem composition and structure, such as reducing the risk of uncharacteristic wildfire effects, will be allowed where access can be gained through existing roads or by equipment not requiring roads ....

The Roadless Area Conservation rule, unlike the establishment of wilderness areas, will allow a multitude of activities including motorized uses, grazing, and oil and gas development that does not require new roads to continue in inventoried roadless areas ....[39]

Certainly, only Congress can designate areas for inclusion in the National Wilderness Preservation System [40]. However, the MUSYA (enacted before the 1964 Wilderness Act), expressly provides for the administrative management of national forest lands for fish and wildlife, outdoor recreation, and watershed purposes, as well as for timber, and states that "the establishment and maintenance of wilderness areas is consistent with [MUSYA's] purposes."[41]. The NFMA (enacted after the Wilderness Act), directs that forest plans "assure ... coordination of outdoor recreation, range, timber, watershed, wildlife and fish, and *wilderness* ...."[42]. Therefore, it appears that, as a general matter, some new prohibitions on activities in roadless areas could lawfully be imposed administratively and the roadless rule might be defended as appropriate management of non-timber resources for multiple use purposes (such as outdoor recreation, mineral development, game and other wildlife), yielding those benefits without permanent impairment of the lands. However, it also is possible that, as applied, restrictions that were severe and extensive might be challenged as violating the "sustained yield" aspects of the MUSYA.

Some of these issues have been raised in suits challenging the roadless area rule, and on July 14, 2003, Judge Brimmer of the Federal District Court for Wyoming permanently enjoined the roadless rule, in part on NEPA grounds and in part on the ground that it was a "thinly veiled attempt to designate 'wilderness areas' in violation of the clear and unambiguous process established by the Wilderness Act for such designation." [43]. The court equated the roadless areas with de facto wilderness, noted the severity of the restrictions under the roadless rule, which the court characterized as restrictive, or more so, than that of congressionally designated wilderness areas, [44] and that the argument that the roadless rule permits multiple uses (such as motorized uses, grazing, and oil and gas development) to proceed so long as roads were not constructed "fails because all of those uses would, in fact, require the construction or use of a road."[45]. Because it felt that the roadless areas under the new rule were tantamount to wilderness areas, the court concluded that the rule was "in violation of the clear and unambiguous process established by the Wilderness Act for such Designation."[46] The court did not reach Wyoming's assertions that the roadless rule also violated NFMA and MUSYA, but did mention that the Wilderness Act "provides protection for a use of the National Forests that was not contemplated by either the Organic Act or the MUSYA ...,"[47] and repeatedly equated roadless areas generally with congressionally designated wilderness.

As discussed above, both MUSYA and NFMA mention wilderness as a use of the national forests, a point the Brimmer opinion did not discuss. Furthermore, in the compromise "release" language in state-by-state wilderness acts since the mid-1980s, Congress repeatedly declined to direct that roadless areas that were not designated as part of the National Wilderness Preservation System when forest management plans were revised be managed for only non-wilderness uses, but instead permitted their management for uses that might maintain their wilderness attributes. The court noted part of the legislative history of the

Wilderness Act to the effect that "the statutory framework of the Wilderness Act would ... assure that no further administrator could arbitrarily or capriciously either abolish wilderness areas that should be retained or make wholesale designations of additional areas in which use would be limited."[48]. The validity of the assertion that the roadless rule was "in violation of the clear and unambiguous process established by the Wilderness Act for such [Wilderness] Designation,"[49] would seem to depend on whether one agrees that management of the roadless areas under the roadless rule would be as restrictive as that under the Wilderness Act, and on how one views the references to wilderness in MUSYA and NFMA and the actions of Congress in continuing to allow management of the roadless areas that maintains wilderness characteristics on national forest lands.

ADMINISTRATIVE ACTIONS AND LITIGATION SINCE JANUARY 20, 2001

The "Card" Memorandum

Immediately after President Bush took office, his Chief of Staff, Andrew Card, issued a memo that directed, among other things, that the effective date of regulations that had been published in the Federal Register, but had not yet taken effect, be postponed for 60 days, unless a department head appointed by President Bush had reviewed and approved the regulatory action [50]. The roadless area regulation was covered by this language, since although it was published as a final rule on January 12, 2001, it was not to be effective until March 13, 2001 [51]. The delay was because the roadless rule was determined to be a "major" rule under the Congressional Review Act, under which Congress is given a certain amount of time to possibly take action to disapprove the rule [52]. If Congress had disapproved the roadless area rule and the President had signed the resulting act, that new legislated direction, of course, would have been binding, but Congress did not take action.

Effective Date Postponed

On February 5, 2001, notice was published in the Federal Register [53] postponing for 60 days the effective date of the roadless area rule from its previous effective date of March 13, 2001, to May 12, 2001 [54]. The Administration then decided to implement the rule, but to consider amending it.

Implementation Enjoined — Part 1

The state of Idaho sued for a declaratory judgment and to enjoin implementation of the roadless rule for violation of NEPA, NFMA and the APA, and other suits in other states also were filed [55]. The court in the Idaho case found that plaintiffs were likely to succeed on their assertion that the FS had not provided the public an opportunity to comment meaningfully on the rule in that there was inadequate identification of the inventoried roadless areas (the court noting that statewide maps were not made available until after the public comment period had ended), inadequate information was presented during the scoping process (FS employees were alleged to be ill-prepared), and the period for public comment was not adequate (all of the public meetings in Idaho occurred within 12 business days of the end of the first 60-day comment period and many of the public comments were received within the last week of the time given and no responses were provided). The court characterized the comment period as "grossly inadequate" and an "obvious violation" of NEPA. The court further found that the FEIS did not consider an adequate range of alternatives, since all but the "no action" alternative included "a total prohibition" on road construction and the EIS did not analyze whether other alternatives might have accomplished protection of the environmental integrity of the roadless areas. In addition, the court concluded that FS did not analyze possible mitigation of negative impacts of the alternatives it did study.

The Bush Administration did not defend the rule, but did ask the court to postpone ruling on the motion for preliminary injunction until it had had an opportunity to complete a full review of the rule, arguing that an injunction was not necessary because the rule was not to be implemented until at least May 12, 2001. The court reserved its ruling until on or after May 4th, the day that the Administration was to submit a status report on its review and findings. On May 4th, the Administration filed its status report with the court and announced that it would implement the Roadless Rule, but would take additional actions to address "reasonable concerns raised about the rule" and ensure implementation in a "responsible common sense manner," including providing greater input at the local planning level [56].

However, on May 10th, Judge Lodge granted a preliminary injunction to prevent implementation both of the Roadless Rule and of the portion of the Planning Rule that relates to prescriptions for the roadless areas (36 C.F.R. § 219.9(b)(8)). The court found the Government's "vague commitment" to propose amendments to the Rule indicative of a failure to take the requisite "hard look" that an EIS is expected to perform, leaving the court with the "firm impression" that implementation of the Roadless Rule would result in irreparable harm to the National Forests. The court concluded that the government's response was a "band-aid approach" and enjoined implementation of the Rule while the agency goes forward with its new study and development of proposed amendments.

The United States did not appeal this decision, but environmental groups who had been granted intervenor status did appeal. Several other lawsuits raising various issues were filed, including suits in North Dakota, Idaho, Alaska and the District of Columbia. The Ninth Circuit reversed the district court and remanded the case to the district court, as will be discussed further below.

Advance Notice of Proposed Rulemaking on Roadless Area Management

On July 10, 2001, the Forest Service published an Advance Notice of Proposed Rulemaking and asked for public comment on ten questions relating to "key principles" involving management of the roadless areas. Comments were due by September 10, 2001, on such questions as: what is the appropriate role of local forest planning in evaluating roadless management; what are the best ways to work collaboratively; how to protect the forests, including protection from severe wildfires; how to protect communities and homes from wildfires on federal lands; how to provide access to nonfederal properties; what factors the FS should consider in evaluating roadless area management; what activities should be expressly prohibited or allowed in roadless areas through the planning process; should roadless areas protected under a forest plan be proposed to Congress for wilderness designation or should they be maintained under a specific roadless management regime; how should the FS work with individuals and groups with strongly competing views; and what other concerns relate to the roadless areas.

Interim Management of Inventoried Roadless Areas I

The final Clinton administrative policy on National Forest System roads published on January 12, 2001, [57] provided interim direction on the management of roadless areas and the construction of roads in roadless areas that was to apply until a roads analysis was completed and incorporated into the relevant forest plans. This direction was in the Forest Service Manual (FSM) and contained considerable detail that would have permitted new roads only if the Regional Forester determined there was a compelling need for the road and both an EIS and a science-based roads analysis had been completed. Examples of instances that constituted compelling need were provided. The management direction was to apply to both inventoried roadless areas and to areas of more than 1,000 acres that were contiguous to inventoried roadless areas (or certain other areas) and met stated criteria. Exceptions were provided to the applicability of the interim guidelines.

Interim Management of Inventoried Roadless Areas II

Pending expected publication of proposed new roadless area rules, the Bush Administration issued a series of Interim Directives affecting roadless area protection and management. The first Directive was effective May 31, 2001, but was not published until August [58]. On June 7, 2001, additional new interim roadless area management direction was provided. On that date, the new Chief of the Forest Service issued a memorandum addressing protection of roadless areas and requiring his approval for some proposed roads or timber harvests in inventoried roadless areas pending completion of forest plan revisions or amendments. Additional directives issued in December of 2001 were difficult to interpret [59]. These ambiguities are relevant because some of these directives were later reinstated, as will be discussed below.

As with earlier directives, the December directive was already in effect (as of December 14, 2001) when published, but retroactive comment was invited — to be considered if final directives were developed. However, the interim directive was only to be in effect for 18 months, unless this time were extended to 36 months, and also apparently was to cease to apply once a forest plan was revised or amended.

As noted, the December directive moved some provisions that more directly address roadless area management into the planning part of the Manual. Only inventoried roadless areas were subject to the interim requirements. The December Directive reserved, as did the earlier ones, authority to the Chief to approve or disapprove certain proposed timber harvests in inventoried roadless areas until a plan revision or amendment was completed "that has considered the protection and management of inventoried roadless areas pursuant to FSM 1920." It also provided that the Chief could designate an Associate Chief, Deputy Chief, or Associate Deputy Chief on a case-by-case basis to be the responsible official. This delegation authority was changed to also include Forest Supervisors in the reinstated and amended directive.

The Regional Forester was to screen timber harvest projects in inventoried roadless areas for possible referral to the Chief. The Chief was to make decisions regarding harvests *except* for those that were: (1) generally of small diameter material, the removal of which is needed for habitat or ecosystem reasons (including reducing fire risk), (2) incidental to a management activity not prohibited under the plan; (3) needed for personal or administrative use; or (4) in a portion of an inventoried roadless area where harvests had previously taken place and the roadless characteristics had been substantially altered. Decisions as to these harvests were to be made by forest officers normally delegated such authority under existing FSM §2404.2. (These delegations included Forest Service line officers).

The December 2001 directive stated that the Chief's authority with respect to timber harvests did not apply if a Record of Decision for a forest plan revision was issued as of July 27, 2001 — as was true of the Tongass National Forest — and would otherwise terminate when a plan revision or amendment that has considered the protection and management of inventoried roadless areas was completed.

The Chief's authority with respect to road construction was to remain in effect until a forest-scale roads analysis was completed and incorporated into each forest plan, at which point it terminated [60]. The Regional Forester was to make many decisions on road construction projects under new §1925.04b. There was no express provision in that section for termination of the authority of the Regional forester. However, the general policy section, 1925.03, keyed termination of the special provisions to completion of a roads analysis and its incorporation into the relevant forest plan:

> Inventoried roadless areas contain important environmental values that warrant protection. Accordingly, until a forest-scale roads analysis (FSM7712.13b) is completed and incorporated into a forest plan, inventoried roadless areas shall, as a general rule, be managed to preserve their roadless characteristics. However, where a line officer determines that an exception may be warranted, the decision to approve a road management activity or timber harvest in these areas is reserved to the Chief or the Regional Forester as provided in FSM 1925.04a and 1925.04b [61].

Under FSM 1925.04a, the Chief had approval authority over all road construction and reconstruction except those decisions delegated to the Regional Forester. Under FSM 1925.04b, the Regional Forester was to screen proposed road projects, forward certain of them to the Chief for approval, but be the deciding officer for many decisions on road projects in inventoried roadless areas, such as when a road is needed for certain purposes.

Note that the December Directive apparently eliminated the requirement that there be a *compelling* need for a road and also eliminated the requirements for a science-based analysis and a full EIS in all cases. In addition, the applicability of the interim direction to certain important contiguous areas also was eliminated. Although the responsible official could still do an EIS and could protect contiguous areas, and a compelling need for a road might exist in some instances, less protection to roadless areas could result because although the new directive *permitted* protection, it did not contain the higher thresholds for approval of activities and more formalized documentation requirements of the previous direction.

This interim directive expired on June 14, 2003. However, as discussed under the heading "2005 Roadless Areas Regulations" below, the directive, renumbered and with certain changes, was later reinstated.

Changes to "Categorical Exclusions"

Changes to agency NEPA documentation requirements also could significantly affect the roadless areas. Under NEPA, agencies must prepare an EIS for proposed actions that might have a significant effect on the human environment. If it is not clear whether if an action might have such an effect, the agency is to prepare an environmental assessment (EA) to determine if an EIS is necessary. Depending on what the EA finds, preparation of an EIS may then follow, or the agency may issue a Finding of No Significant Impact (FONSI), in which case no further analysis is required. However, some actions have been shown to have so little effect on the environment that not even an EA is necessary. An agency may indicate what these clearly non-harmful actions are through its articulation of "categorical exclusions" — actions that are excluded from preparation of even an EA [62].

Several agency actions have expanded the activities that may be conducted as categorical exclusions, and have replaced previous categorical exclusions on timber sales. Joint Forest Service/Bureau of Land Management categorical exclusions were finalized for hazardous fuels reduction projects that can include mechanical treatments on up to 1,000 acres [63]. A categorical exclusion was also finalized for smaller FS timber sales [64]. In addition, certain silvicultural treatments may proceed as categorical exclusions under Title IV of the Healthy Forests Restoration Act [65].

Previously, the categorical exclusions portion of the FS Handbook set out types of activities that normally would be excluded from NEPA documents — unless extraordinary circumstances are present. One of the listed extraordinary circumstances was the presence of inventoried roadless areas. Extraordinary circumstances were defined as "conditions associated with a normally excluded action that are identified during scoping as *potentially* having effects which *may* significantly affect the environment." (Emphasis added.) The presence of an extraordinary circumstance arguably removed the proposed action from qualifying as a categorical exclusion and required the preparation of an EA in order to probe further the possible environmental effects. This is the interpretation of the Handbook section and its legislative history in a Seventh Circuit case [66].

New interim guidance was proposed [67] and finalized, [68] such that the presence of circumstances previously considered "extraordinary" do not necessarily preclude an action from being a categorical exclusion if the responsible official, based on scoping, determines there would be no significant environmental effects. Indeed, under the new directive, a circumstance *is* "extraordinary" only if the responsible official determines it is because the proposed action may have a significant effect on the environment, in which case an EIS is to be prepared. Or, if the responsible official is uncertain whether the proposed action may have a significant effect on the environment, an EA is to be prepared. Under this approach, the presence of circumstances that previously were per se "extraordinary" now are merely "resource conditions" that must specifically be considered in determining whether an action may have a significant effect. "It is the degree of the potential effect of a proposed action on these resource conditions that determines whether extraordinary circumstances exist." [69]/ This is a significant change from the previous text.

In defense of the proposed changes on explanatory circumstances, the explanatory material asserted that there is a split in the decisions of the circuits on the effects of the presence of extraordinary circumstances, and that the Ninth Circuit has held that an agency may issue a categorical exclusion even where a certain resource condition, such as the presence of threatened or endangered species, is found [70]. However, the cited case involved a salvage sale under §2001 of the Rescissions Act, [71] a statute that sets out a very narrow scope of judicial review of environmental decisions and a very broad range of discretion in the Secretary to determine the adequacy of any environmental reviews. In contrast, the Seventh Circuit opinion, which analyzed the wording and derivation of the current categorical exclusion provisions was not so contextually limited, and hence is arguably more on point [72]. Under the final directive, some timber sales could be conducted as categorical exclusions in roadless areas (and possibly in a roadless area with endangered or threatened species) if the official determines, without the necessity of written documentation of the underlying analysis relied upon, that there would be no significant environmental effects.

Comments on Possible Roadless Changes

On June 26, 2002, the Forest Service released its summary report dated May 31, 2002, on the public comments received in response to the Advance Notice of Proposed Rulemaking. The Forest Service received approximately 726,000 responses, said to be mostly form letters to the 10 questions, but which included 52,432 original responses. The report includes appendices that describe the system used to analyze the comments, and urges caution in relying on the gist of the comments received, in that "respondents are self-selected; therefore their comments do not necessarily represent the sentiments of the entire population. The analysis attempts to provide fair representation of the wide range of views submitted, but makes no attempt to treat input as if it were a vote." Appendix E indicates that the overwhelming number of "organized" responses were in favor of the Roadless Rule [73].

Ninth Circuit Decision on 1st Injunction

On December 12, 2002, [74] the Ninth Circuit reversed the Idaho district court stating:

> We hold that the district court had discretion to permit intervention, under Fed. R. Civ. P. 24(b), and intervenors now can bring this appeal under Fed. R. Civ. P. 24(b); that plaintiffs have standing to challenge the Roadless Rule; and, assessing the merits, that the district court abused its discretion in granting [sic] preliminary injunction against implementation of the Roadless Rule [75] . . . . Because of its incorrect legal conclusion on prospects of success, the district court proceeded on an incorrect legal premise, applied the wrong standard for injunction, and abused its discretion in issuing a preliminary injunction [76]

Idaho's petitions for panel rehearing was denied on April 4, 2003.

In reaching its conclusions, the Ninth Circuit reviewed the substantive grounds considered by the district court and disagreed that plaintiffs had demonstrated a likelihood of success on the merits, finding instead that the Forest Service did adequately comply with NEPA in its provision for public comment on the Roadless Rule because the maps provided did not suffer from the grave inadequacies alleged by plaintiffs, plaintiffs had actual notice as to the roadless areas that would be affected, and, at most, possibly inadequate maps would only affect the propriety of the Rule on the 4.2 million acres added during the EIS process. The court also found that the Forest Service had provided more than the minimum required amount of time for comment, the time allowed was adequate, [77] and that the EIS considered an adequate range of alternatives [78]. Because it felt that the district court wrongfully found that plaintiffs were likely to succeed on the merits, the appellate court concluded that the district court accepted only a minimal showing of irreparable harm and incorrectly issued the injunction. the Ninth Circuit denied Idaho's petition for rehearing, and the case was remanded to the district court to reconsider its previous reasoning and injunction in light of the opinion of the appellate court. Several other lawsuits challenging the Roadless rule were stayed pending the decision by the Ninth Circuit. Because litigation in the 10th Circuit was dismissed as moot, the opinion of the Ninth Circuit is the only precedent on these issues.

Most of the unfavorable response to the Ninth Circuit decision has focused on whether it was proper for the intervenors to bring the appeal when the government did not. It will be recalled that the case came forward in an unusual context: although several statutes were initially involved in the lawsuits, the district court decision focused on the inadequacy of the federal defendants' NEPA compliance, a decision the federal defendants did not appeal. Certain environmental groups had been granted intervenor status and appealed the district court's ruling. The decision of the Ninth Circuit raised significant issues relating to whether the intervenor groups could appeal NEPA-compliance rulings when the federal defendants — the only ones who could comply with NEPA — did not [79]

Interim Implementation; Tongass

It will be recalled that the Clinton Administration on January 12, 2001, had issued interim direction for the management of roadless areas until a roads analysis had been completed and incorporated into relevant forest plans. This interim direction was followed by other interim directives issued by the Bush Administration.

On June 9, 2003, the Bush Administration indicated that it would "retain" the roadless rule and would not renew the interim management directives. However, the roadless rule would be modified to exclude the Tongass National Forest as part of a settlement of the lawsuit filed by the State of Alaska. A proposed rule to exclude the Tongass from the broader protection of the roadless rule was published on July 15 [80] and an Advance Notice of Proposed Rulemaking to exclude both the Tongass and the Chugach National Forest (also in Alaska) was also published on the same day [81]. A final rule "temporarily" exempting the Tongass from the roadless rule was published on December 30, 2003 [82]. Specifically, the roadless area restrictions on road construction, road reconstruction, or the cutting, sale, or removal of timber in inventoried roadless areas would not apply on the Tongass National Forest. The explanatory material indicated that the exemption was temporary — until an Alaska-wide roadless rule was finalized. The exemption meant that roadless areas protected under the Revised Land Management Plan for the Tongass would remain protected, but an additional 300,000 acres of roadless areas would be available for logging under that plan.

The Forest Service has taken the position that the finalization of the new roadless rule on May 13, 2005 (discussed below), has eliminated the need for further Tongass-specific rulemaking and that timber harvest decisions that include a roadless component will be made in accordance with the forest plan unless changed through state-specific rulemaking [83].

Implementation Enjoined — Part 2

On July 14, 2003, Judge Brimmer of the Federal District Court for Wyoming permanently enjoined the Roadless Rule, in part because of NEPA defects, and in part because he concluded it created de facto wilderness in violation of the Wilderness Act. [84] (See discussion of the wilderness issues, *supra.*) The court concluded that 1) the FS's decision not to extend the scoping comment period was arbitrary and capricious; 2) the FS's denial of cooperating agency status to Wyoming and other states was arbitrary and capricious; 3) the FS's failure to rigorously explore and objectively evaluate reasonable alternatives to the Roadless Rule was contrary to law; 4) the FS's failure to adequately analyze cumulative effects was a clear error in judgment; and 5) the FS's decision not to issue a supplemental EIS in light of new information on updated roadless area inventories was arbitrary and capricious, and contrary to law [85]. The court stated that the agency drove the rule "through the administrative process in a vehicle smelling of political prestidigitation" [86] in its "rush to give President Clinton lasting notoriety in the annals of environmentalism"[87] —and concluded that the agency must "start over" [88]. On July 11, 2005, the Tenth Circuit dismissed the appeal in this case and vacated the district court's decision. It held that new roadless regulations published on May 13, 2005, discussed below, made the case moot [89].

New Proposed Roadless Regulations

On July 16, 2004, a new proposed roadless rule was published [90]. Under the proposed changes to 36 C.F.R. §394, no protections of roadless areas would be prescribed in the rule. Instead, a new procedure would be put in place under which there would be a window of time within which the governor of a state could petition the Secretary of Agriculture to promulgate a state-specific rule on the management of roadless areas.

Interim Management of Inventoried Roadless Areas III

Effective upon publication of the proposed rules in the Federal Register, and for the duration of the time allowed for the state petition process, the FS reinstated the December 2001 directive 1920-2001-1, renumbered as ID 1920-2004-1 and modified it in two respects [91]. This reinstatement put in place again the provisions described above that allow the Chief of the FS to make decisions affecting inventoried roadless areas involving 1) road construction — until a forest-scale roads analysis is completed and incorporated into a forest plan or a determination is made that a plan amendment is not needed; and 2) the cutting of timber until a forest plan is in place that "considers" the protection and management of roadless areas.

ID 1920-2004-1 differs from the previous one in that it allows the Chief to grant project-specific exceptions to allow a Regional Forester or a Forest Supervisor, for good cause, to exercise the authority to conduct projects in roadless areas. The addition of forest supervisors to those who can conduct projects in roadless areas is an expansion of the previous ID that arguably could make it easier to approve projects in those areas. Secondly, FSM 1925.04b is changed to allow the Regional Forester to make decisions on road construction and reconstruction in an inventoried roadless area for lands associated with any mineral lease, license, permit or approval for mineral leasing operations.

NEW FOREST PLANNING REGULATIONS

New regulations on National Forest System land management planning were published on January 5, 2005. [92]. A separate rule repealed the 2000 planning regulations at 36 C.F.R. § 219, Subpart A, but did not repeal the roadless area management rule at 36 C.F.R. § 294, Subpart B.

Section 219.7(a)(5)(ii) of the new rules basically retains a provision from the 2000 rules [93] and provides that

> Unless otherwise provided by law, all National Forest System lands possessing wilderness characteristics must be considered for recommendation as potential wilderness areas during plan development or revision.

Depending on how 'wilderness' characteristics and evaluation criteria might be defined, this provision could require many roadless areas to be reviewed for possible wilderness designation as part of the cyclical plan revision process. No management guidance is provided for inventoried roadless areas that are not recommended for inclusion in the National Wilderness Preservation System [94]. Another section indicates that, initially at least, all forest lands generally will be considered to be available for multiple uses going into the planning process [95]. "Roadless area" and "inventoried roadless area" are removed from the definitions section because the terms are not used in the rule. Comments raising concerns about protection for roadless areas were submitted when the planning regulations were proposed and the response was that the "Responsible Official considers these values during the planning process for the plan's desired conditions and objectives, which will involve local, regional, and national interests, and use the best available local information." This response seems to indicate that roadless values will be only one factor among many to be considered in managing the forests [96].

The new planning regulations stress that plans are not decisional in nature, but rather are aspirational only, setting out goals for the management of forest units. Nor do they contain much detail on how to manage the forests, leaving that to the Forest Service Manual, Handbooks, and directives. In particular, the explanatory material states that guidance about special area concerns, such as roadless areas are more properly included in Forest Service directives [97]. In response to another comment that special protection for roadless areas was lost in the new regulations, the materials point only to the provision on wilderness evaluations quoted above, and state that additional guidance on roadless *wilderness evaluations* will be forthcoming in directives [98].

Several publications of interim directives with solicitation of comments occurred since the publication of the final planning rules, principally the 12 IDs published on March 23, 2005 [99]. However, as has been true of previous directives, it is difficult to find the texts in question [100] and to ascertain the current status of a particular management issue — in this case the management of inventoried roadless areas. In addition, the relationship of the March 23rd ID to the previous interim directive on roadless area management is not clear. It may be recalled that ID 1920-2001-1 (issued December 14, 2001, and expired June 14, 2003) was reinstated and renumbered as ID 1920-2004-1 "to provide guidance for addressing road and timber management activities in inventoried roadless areas until land and resource management plans are amended or revised."[101] (Again, although that July 16, 2004, publication indicates the ID is available at [http://www.fs.fed.us/im/directives], it in fact is not — but see the discussion of it in the preceding section of this report.) The March 23rd interim directives include ID 1920-2005-1, which states that it does not supersede any previous IDs. The new 1920 series directive does not address the previous directive bearing that number, nor address roadless area management at all, except to indicate that part 1925 of the Forest Service Manual is "reserved" for IDs and field management guidance on the Management of Inventoried Roadless Areas — with no indication that there is in fact already such a directive in effect that is not referenced. Therefore, it was not clear whether there was any direction in effect with regard to management of roadless such areas pending revision of plans under the new planning regulations.

The final 2005 roadless area rules discussed below originally stated that the July 16, 2004, interim directive would remain in place until January 16, 2006, unless renewed for an additional 18 months [102]. The directive was so renewed on January 16, 2006, and now is scheduled to remain in place until July 16, 2007, or until a statewide rule is finalized under the new roadless area regulations or a forest plan "considers" roadless areas [103].

2005 ROADLESS AREA REGULATIONS

New final roadless area rules were published on May 13, 2005 [104]. The final roadless area regulations are very similar to the proposed regulations. The Forest Service asserts that the rule incorporates the department's five conservation principles for inventoried roadless areas:

1. Make informed decisions to ensure that inventoried roadless area management is implemented with reliable information and accurate mapping, including local expertise and experience.
2. Work with states, tribes, local communities and the public through a process that is fair, open, and responsive to local input and information.
3. Protect forests to ensure that the potential negative effects of severe wildfire, insect, and disease activity are addressed.
4. Protect communities, homes and property from the risk of severe wildfire and other risks on adjacent federal lands.

5. Ensure that states, tribes, and private citizens who own property within inventoried roadless areas have access to their property as required by existing law [105]

The new regulations establish a procedure to permit a Governor of a state to submit a petition to the Secretary regarding roadless areas in a state, and to make recommendations as to management of such areas. Petitions are required to provide specified information, including 1) a description of the particular lands for which the petition is made; 2) particular management recommendations; 3) identification of the circumstances and needs intended to be addressed by the petition, including conserving roadless value, protecting health and safety, reducing hazardous fuels, maintaining dams, providing access, etc.; 4) information on how the recommendations differ from the relevant federal plans and policies; 5) how the recommendations compare to state land policies and management direction; 6) how the recommendations would affect fish and wildlife and habitat; 7) a description of any public involvement efforts undertaken by the state during development of the petition; and 8) a commitment by the state that it will participate as a cooperating agency in any environmental analysis for the rulemaking.

Petitions must be submitted not later than November 13, 2006, but the explanatory material mentions that petitions for rulemaking or amendment after that time could be filed under 7 C.F.R. § 1.28. If the Secretary approves a petition, the FS must coordinate development of the proposed rule with the state, but the Secretary or the Secretary's designee is to make the final decision on any rule.

The rationale for imposing a deadline on the roadless area special rule process, and the consequences of a state's failure to meet the deadline, are unclear. The rule does not indicate what process will pertain if a newly-seated Governor of a state that previously failed to submit a petition under the roadless rule petitions under 7 C.F.R. § 1.28 for a state-specific roadless rule. Section 1.28 states only that petitions filed under the section "will be given prompt consideration and petitioners will be notified promptly of the disposition made of their petitions." Concerns relating to this ambiguity are relevant, as it has been reported that the Governors of Minnesota, New Hampshire, Utah, and Wyoming have decided not to submit petitions for state-specific roadless rules in accordance with the deadline [106]

The rule does not include any standards on the scope or balance required for public participation in the development of a state's recommendations, or for how a state is to gather its information about an area's resources and values. There also are no requirements or standards for the Secretary's review and approval of a petition, and, perhaps most importantly, no statement as to the relative weight to be given to a state's recommendations versus the preferences of non-state residents who may express an opinion on the desired management of these national lands [107].

The background materials accompanying both the proposed and final rule refer to the importance of collaborating with partners, including state governments, stating: "strong State and Federal cooperation regarding management of inventoried roadless areas can facilitate long-term, community-oriented solutions." [108]. The materials review the promulgation of the original roadless rule and note that "concerns were immediately expressed by those most impacted by the roadless rule's prohibitions" and that the new rule is proposed in response to those concerns [109]. This latter comment raises interesting questions about the role of public input in the rulemaking in that only approximately 4% of the extraordinary number of comments received on the original roadless rule opposed or expressed concerns about its protections. Similarly, the overwhelming number of responses to the new roadless rule favored continuing protection. The explanatory materials justify taking a different approach by stating that "every comment received is considered for its substance and contribution to informed decisionmaking, whether it is one comment repeated by tens of thousands of people or a comment submitted by only one person. The public comment process is not intended to serve as a scientifically valid survey process to determine public opinion"[110].

Advisory Committee Established

In accordance with the new roadless rule, a Roadless Area Conservation National Advisory Committee was established by the Secretary under the Federal Advisory Committee Act, to provide advice and recommendations on the implementation of the state petitions for special roadless area management rules [111]. The Committee was to consist of up to 15 members representing a "balanced group of representatives of diverse national organizations who can provide insights into the major contemporary issues associated with the conservation and management of inventoried roadless areas" [112]. The Committee was to operate "in a manner designed to establish a consensus of opinion." On December 14, 2005, the Committee held its first meeting in Washington, D.C [113].

Multi-State Lawsuit to Enjoin 2005 Roadless Area Rule

On August 28, 2005, the States of California, Oregon, and New Mexico filed a lawsuit in the Federal District Court for California to challenge the 2005 roadless area rule [114]. The State of Washington later joined the suit as a plaintiff intervenor. The Earthjustice Legal Defense Fund and other environmental organizations also filed suit to challenge the rule, and the two cases have been consolidated [115]. The plaintiffs allege that the Department of Agriculture violated NEPA and the Administrative Procedure Act, 5 U.S.C. §701 *et seq.* (APA), when it promulgated the 2005 roadless area regulations. They argue that the Department finalized the regulations without conducting requisite NEPA environmental analysis, allowing meaningful public involvement, or explaining the rationale for the Department's change in roadless areas policy [116]. The case currently is pending.

Initial State Responses to 2005 Roadless Area Rule

Several governors have expressed concern that the new roadless area rule petition process is too vague to provide states proper guidance in proposing state-based rules and places undue burdens on state resources. North Carolina Governor Michael Easley raised the vagueness issue in a letter to Secretary of Agriculture Ann Veneman, asserting that the rule provides "insufficient clarity" on how the Department will respond to state petitions or the criteria the Department will use to consider them [117]. The Governor also noted that the rule does not clarify how the Department will coordinate state-specific rule development among individual states or how it will approach management of roadless areas that cross state boundaries [118].

Governors also have expressed concern over the rule's burden on states. On October 14, 2004, Oregon Governor Theodore Kulongoski filed a petition with Secretary Johanns to permit states to adopt the 2001 roadless area rule through an expedited process. The Governor argued that the 2005 roadless area rule will require states to expend significant financial and personnel resources to ensure adequate public participation and technical analysis relating to its provisions and implementation [119]. North Carolina Governor Easley stated in his August 2004 letter to Secretary Venemen that the administrative and financial investment in the petitioning process could be "onerous" to state agencies. Tennessee Governor Bredesen similarly stated in a September 8, 2004, letter to Secretary Veneman that the rule may place an "undue financial burden on our state agencies without providing some assurances of any concurrence or approval from the Department of Agriculture." [120]. Virginia Governor Mark Warner expressed concern in a July 30, 2004, letter to Secretary Veneman that, although the 2005 rule confirms that the U.S. Forest Service has final authority regarding management of roadless areas, the rule places on individual states the burden of seeking protection of the areas [121].

State Petition Submissions and Responses

To date, governors of five states have filed petitions with the Secretary of Agriculture under the 2005 roadless area rule. Virginia Governor Warner filed a petition on December 22, 2005, requesting protection for all roadless areas in the state. Subsequently, the Governors of North Carolina, South Carolina, New Mexico, and California submitted similar petitions requesting comprehensive protection of all roadless areas in their respective states.

The Advisory Committee recommended that the petitions of North Carolina, South Carolina, and Virginia be approved, and on June 21, 2006, the Department of Agriculture approved them [122]. Letters from the Department to the governors outlined the process for moving forward with proposed state-specific roadless area rules —that the Department and the respective state should take the following steps: develop a memorandum of understanding regarding the "appropriate level of environmental analysis" required under NEPA and the Endangered Species Act; ensure that the rule provides for public involvement and "active solicitation" of the views of interested parties; coordinate the development of the rule; and consider how the rule will amend any existing state land management plans [123]. The letters reiterated that the proposed state-based rules would be subject to federal notice and comment rulemaking requirements and that the Secretary has sole authority to formally adopt any such rules after issues emerging from the notice and comment period have been resolved.

No decision on the petition of New Mexico has been made by the advisory committee.

Pending Legislative Responses

Several bills have been introduced in the 109th Congress in response to the roadless area initiatives of the Bush Administration. On October 19, 2005, then Senator Jon Corzine of New Jersey introduced S. 1897, The Act to Save America's Forests [124]. This measure would provide more comprehensive protections of roadless areas than did the 2001 rule. On March 2, 2006, Senator Marie Cantwell of Washington introduced S. 2364, The Roadless

Area Conservation Act of 2006 [125]. The measure essentially would enact the previous 2001 rule.

CONCLUSION

Roadless areas of the National Forest System have received special management for decades and more recently have been addressed by two administrations through a complex series of interrelated actions on roads, roadless areas, and forest planning. A new nationwide rule providing protections and use restrictions on inventoried roadless areas was promulgated in January 12, 2001, in an effort to standardize management of such areas and avoid persistent litigation. However, a federal court enjoined implementation of the Clinton Administration's roadless rule, and although this decision was later reversed, another federal district court has again permanently enjoined implementation of the rule. A new roadless area rule has been finalized that replaces previous protections with a new process by which a governor of a state may petition the Secretary to promulgate a state-specific rule for the management of roadless areas. A petition would also contain recommendations for the management of the areas. The proposed rule does not contain details on how the public is to be involved in the development of a state petition, or how the state's recommendations are to relate to the preferences expressed by non-state residents regarding the same public lands. Virginia, North Carolina, South Carolina, New Mexico, and California have submitted petitions so far, and the first three of these have been approved and will proceed to development of state-specific rules.

During the time allowed for the new process, previous interim direction for the management of the roadless areas was reinstated, with some modifications, to govern the management of these areas during the time allowed for the new process. This direction allows the Chief of the FS to make decisions as to roads and timber cutting, but to delegate that responsibility in some circumstances, including to authorize forest supervisors to conduct projects in roadless areas. The interim provisions currently are in effect until July 16, 2007, a statewide rule is finalized, or a forest plan "considers" roadless areas.

New final planning regulations published on January 5, 2005, appear to presume that roadless areas are basically available for a variety of uses, including timber harvests, subject to unit-by-unit planning processes. Other administrative changes regarding categorical exclusions and extraordinary circumstances now appear to allow certain actions to be taken in roadless areas without written environmental analyses.

In summary, the regulatory direction for the management of the remaining National Forest System inventoried roadless areas arguably has gone from embodying a special recognized status with nationwide protections to having no special status or even separate mention in the new forest-by-forest planning regulations. Also, a new petition process has been established under which a Governor of a state may seek a special statewide rule regarding roadless areas. What management protections, if any, may apply to roadless areas in the future is not clear.

REFERENCES

[1]   The NFS includes the national forests and national grasslands and is administered by
        the Forest Service in the Department of Agriculture. Roadless areas within the NFS
        have long received special management. Beginning even before enactment of the 1964
        Wilderness Act, larger (generally 5,000 acres or more) roadless areas were
        "inventoried" to consider their wilderness characteristics, and later their suitability for
        inclusion in the National Wilderness Preservation System. These are the "inventoried"
        areas referred to in the Administration's initiative and in this report. A discussion of the
        roadless area initiative and many of the related documents are available on the Forest
        Service website at [http://www.roadless.fs.fed.us/].
[2]   70 Fed. Reg. 25,654.
[3]   Memorandum from President William J. Clinton to the Secretary of Agriculture on
        Protection of Forest 'Roadless' Areas, Oct. 13, 1999.
[4]   64 Fed. Reg. 56,306.
[5]   65 Fed. Reg. 30,276.
[6]   66 Fed Reg. 3,244 (January 12, 2001), adding 36 C.F.R. § 294, Subpart B..
[7]   Id., at 3,246.
[8]   63 Fed. Reg. 4,350, regarding regulations at 36 C.F.R. § 212.
[9]   63 Fed. Reg. 4,354.
[10]   64 Fed. Reg. 7,290 (Feb. 12, 1999).
[11]   65 Fed. Reg 11,680 (Mar. 3, 2000).
[12]   65 Fed. Reg. 11,684.
[13]   66 Fed. Reg 3,206 and 3,219 respectively.
[14]   "Forest development roads" is changed to "National Forest System roads" and "forest
        transportation plan" is changed to "forest transportation atlas." Other new definitions
        also are added, e.g. to clarify "road construction" and "road reconstruction." 66 Fed.
        Reg. 3,216-3,217.
[15]   66 Fed. Reg. 3,230.
[16]   65 Fed. Reg. 67,514. Revising the planning regulations has been a contentious issue for
        the Forest Service for quite some time.
[17]   The compliance date was extended in an interim final rule, 66 Fed. Reg. 27,552 (May
        17, 2001). Compliance with the 2000 regulations was later postponed until new
        planning regulations are finalized. 67 Fed. Reg. 35,431 (May 20, 2002). See also
        extension of compliance for projects implementing plans, 68 Fed. Reg. 53,294
        (September 10, 2003).
[18]   70 Fed. Reg. 1,023.
[19]   Act of June 4, 1897, ch. 2, 30 Stat. 34.
[20]   P.L. 86-517, 74 Stat. 215.
[21]   P.L. 94-588, 90 Stat. 2949, primarily amending P.L. 93-378.
[22]   16 U.S.C. § 475.
[23]   16 U.S.C. § 551.
[24]   16 U.S.C. § 531.
[25]   16 U.S.C. § 528.
[26]   16 U.S.C. § 529.

[27] P.L. 88-577, 78 Stat. 890.

[28] 16 U.S.C. § 531.

[29] 16 U.S.C. § 1604(g). Note that "wilderness" management is again mentioned, twelve years after enactment of the Wilderness Act.

[30] P.L. 88-577, *supra.*

[31] 66 Fed. Reg. 3,245 (Jan. 12, 2001).

[32] *Id.,* at 3,246.

[33] New 36 C.F.R. § 294.13(b)(1), 66 Fed. Reg. 3,273.

[34] *Id.,* § 294.13(b)(2)-(4).

[35] *Id.,* § 294.14(a) and (c).

[36] *Id.,* § 294.14(d).

[37] The *proposed* rule would not have applied the prohibitions on new road construction to the Tongass National Forest in Alaska. Rather, decisions on whether the prohibitions should apply to any or all of the inventoried roadless areas in the Tongass would have been considered at the time of the five-year review of the April 1999 revised Tongass Plan (i.e. in 2004). In contrast, the preferred alternative in the FEIS would have applied the road and timber prohibitions to the Tongass in April 2004.

[38] 65 Fed. Reg. 67,529.

[39] 66 Fed. Reg. 3,249.

[40] P.L. 88-577, 78 Stat. 890.

[41] P.L. 86-517, 74 Stat. 215, 16 U.S.C. § 529.

[42] 16 U.S.C. § 1604(e).

[43] Wyoming v. U.S. Department of Agriculture, 277 F. Supp. 2d 1197, 1239 (D. Wyo 2003).

[44] The court reviewed the exceptions by which roads could be allowed in roadless areas under the rule, but did not mention that Federal Aid Highways could be permitted in some instances (p. 1236). On the other hand, the court noted that the roadless rule was more constrained with respect to constructing roads in roadless areas in order to combat problem conditions than was the Wilderness Act, in that the Wilderness Act allows roads "to control fire, insects, and diseases," while the roadless rule only allows roads in the case of an "imminent flood, fire, or other catastrophic event that, without intervention, would cause the loss of life or property." (Comparing 16 U.S.C. 1133(d)(1) with 36 C.F.R. § 294.12(b)(1).

[45] *Id.* .

[46] *Id.* at 1239.

[47] *Id.,* at 1234.

[48] *Id.,* at 1233, quoting from H.Rept. 88-1538.

[49] *Id.,* at 1239.

[50] Memorandum for the Heads and Acting Heads of Executive Departments and Agencies, Andrew H. Card, Jr. January 20, 2001. Exceptions are also made for rules that are subject to statutory or judicial deadlines, or rules the Office of Management and Budget Director deems are excepted because they are needed for an emergency or other urgent situation relating to health and safety.

[51] Several dates surround the roadless rule: the 60-day delayed effectiveness date in the rule itself — which derives from the Congressional Review Act (CRA)(Subtitle E of the Small Business Regulatory Enforcement Act of 1996, P.L. No 104-121, 110 Stat.

857-874, 5 U.S.C. §§801 *et seq.*); the 60-day delay resulting from the President's directive; and the usual 30-day delay that might otherwise apply under the Administrative Procedure Act (APA)(5 U.S.C. 501 *et seq.*). Normally, the 30-day APA delay period and the 60-day CRA delay period run concurrently.

[52] Under 5 U.S.C. § 804(2), a major rule is one that "has resulted in or is likely to result in — (A) an annual effect on the economy of $100,000,000 or more; (B) a major increase in costs or prices for consumers, individual industries, Federal, State, or local government agencies, or geographic regions; or (C) significant adverse effects on competition, employment, investment, productivity, innovation, or on the ability of United States-based enterprises to compete with foreign-based enterprises in domestic and export markets" other than rules under the Telecommunications Act of 1996. Under the Review Act, a rule that has been determined to be a major rule cannot become effective for at least 60 days after publication. This delay period is to give the Congress time to consider the rule and to address legislatively issues raised by it. A major rule will take effect the later of the date occurring 60 days after the date on which — (1) Congress receives the report submitted pursuant to § 801(a)(1); or after the rule is published in the Federal Register, if it is; (2) if the Congress passes a joint resolution of disapproval which is subsequently vetoed by the President, then the earlier of when one House votes and fails to override the veto, or 30 session days after Congress receives the veto message from the President; or (3) the date the rule would otherwise have taken effect if not for the review requirement. 5 U.S.C. § 801(a)(3). Other provisions allow a major rule to become effective earlier under certain circumstances, such as those involving an imminent threat to health or safety or other emergency circumstances, national security, etc., or if either House votes to reject a joint resolution of disapproval. When, as in this instance, a rule is published and/or reported within 60 session days of adjournment of the Senate or 60 legislative days of adjournment of the House through the date on which the same or succeeding Congress first convenes its next session, that Congress may consider and pass a joint resolution of disapproval during a period of 60 session or legislative days after receiving the reported rule. A held-over rule takes effect as otherwise provided; but the opportunity for Congress to consider and disapprove it is extended. The *usual* effective date of a regulation under the APA is 30 days after publication, during which time affected persons may prepare for and adjust to the impending effects of the rule. The 30-day period is intended as a minimum, and an agency may set a longer interval if that appears advisable, and longer times have been afforded in circumstances when it is anticipated that economic adjustments must be made in response to the new regulatory requirements. However, there are exceptions to the applicability of the APA, one of which is that the usual rulemaking procedures do not apply to rules relating to agency management or public property. However, in 1971, Secretary of Agriculture Hardin partially waived the APA exemption for rules related to public property (36 Fed. Reg. 13804 (July 24, 1971). The Hardin Order subjects Department of Agriculture rulemaking to the public notice and comment requirements prescribed by 5 U.S.C. 553(b) and (c), subject to exceptions for good cause. However, the Order does not appear to subject Department rules to the 30-day delay requirement of § 553(d), or to other APA provisions beyond § 553(b) and (c), a fact which may be relevant to options available to change the rule. Many Forest Service rules, including the new Planning rule and the Roads rule, are effective

immediately upon being finalized. Under the APA, interested persons have the right to petition for issuance, amendment, or repeal of a rule, even during the 30-day delay period, although by what procedures an agency may accomplish changes in response to such a petition during the delay period is not totally clear. As noted above, however, the roadless rule does not appear subject to these provisions. Even if it were, the roadless rule is a final published rule, even though it is not yet in effect, and at least one court has held that an agency cannot simply "repeal" such a regulation, but rather may need to modify or revoke the final regulation through commensurate procedures (Consumer Energy Council of America v. Federal Energy Regulatory commission, 673 F. 2d 425 (D.C. Cir. 1982)) — here those that may be required by the NFMA and other agency regulations.

[53] 66 Fed. Reg. 8,899.

[54] The postponement notice stated that the action was exempt from notice and comment either because it is a procedural rule or for good cause shown: "[t]o the extent that 5 U.S.C. section 553 applies to this action, it is exempt from notice and comment because it constitutes a rule of procedure under 5 U.S.C. section 553(b)(A). Alternatively, the Department's implementation of this rule without opportunity for public comment, effective immediately upon publication today in the Federal Register, is based on the good cause exceptions in 5 U.S.C. section 553(b)(B) and 553(d)(3). Seeking public comment is impracticable, unnecessary and contrary to the public interest. The temporary 60-day delay in effective date is necessary to give Department officials the opportunity for further review and consideration of new regulations, consistent with the Assistant to the President's memorandum of Jan. 20, 2001. Given the imminence of the effective date, seeking prior public comment on this temporary delay would have been impractical, as well as contrary to the public interest in the orderly promulgation and implementation of regulations. The imminence of the effective date is also good cause for making this rule effective immediately upon publication."

[55] Idaho v. Dombeck, CV01-11-N-EJL (D.C. Id. 2001); Kootenai Tribe of Idaho et al v. Dombeck, CV01-10-N-EJL. (D.C. Id. 2001) Colorado and Alaska have joined Idaho in the suit and Utah also has filed suit.

[56] USDA News Release No. 0075.01.

[57] 66 Fed. Reg. 3,219.

[58] The first of these (I.D. No. 7710-2001-1) was actually published on August 24, two days after the second of these directives (I.D. No. 7710-2001-2 and I.D. No. 2400-2001-3, both issued July 27, 2001), even though the first one had been in effect since May 31. See 66 Fed. Reg. 44,590 (August 24, 2001) and 66 Fed. Reg. 44,111 (August 22, 2001).

[59] Interim Directive 1920-2001-1 and I.D. No. 7710-2001-3, both effective December 14, 2001, were published on December 20, 2001 (66 Fed. Reg. 65,796). I.D. 1920-2001-1 appeared to substantially replace much of the previous directives. However, the Notice did not clearly indicate which provisions were being replaced or the precise extent of revisions. The published explanatory material stated that affected material was set out and unaffected material was not. Yet some of the earlier provisions were neither shown nor discussed and therefore, might still have been in effect. However, the final text of new FSM §1925 did not show these undiscussed earlier provisions — as though they were superseded. Therefore, it was not clear which of the previous materials were still

in effect. For example, some of former FSM §7712.16 (that contained many specific details on permissible road construction) was expressly revised in the December Directives (notably the former requirements for protection of contiguous areas and the requirement for preparation of an EIS for projects in roadless areas are eliminated) and the explanatory materials stated that the revised provisions were then moved to appear in the Planning part of the Manual as new §1925. Yet other provisions that were in §7712.16 were neither discussed as superseded or modified, nor set out in new §1925. One example is that the previous requirement for a "compelling need" for the road project disappeared without comment. Another example is §7712.16b, paragraph 3, which distinguished between classified and unclassified forest roads and stated that environmental mitigation and environmental restoration of unclassified roads are appropriate in inventoried roadless and contiguous unroaded areas and must follow NEPA-based decisionmaking processes. However, reconstruction or maintenance of unclassified roads in inventoried roadless and contiguous unroaded areas is inappropriate, other than to prevent or correct resource damage, as such activity would lead to de facto road development. These ambiguities made analysis of and comment on the December Directives difficult.

[60] *Id.,* at 65,801.

[61] 66 Fed. Reg. 65,801.

[62] 40 C.F.R. 1508.4.

[63] 68 Fed. Reg. 33,814 (June 5, 2003).

[64] 68 Fed. Reg. 44,598 (July 29, 2003).

[65] P.L. 108-148, 117 Stat. 1887.

[66] Rhodes v. Johnson, 153 F. 3d 785 (7th Cir. 1998).

[67] 66 Fed. Reg. 48412 (September 20, 2001).

[68] 67 Fed. Reg. 54622 (August 23, 2002).

[69] FSH 19909.15 — Environmental Policy and Procedures Handbook, § 30.3b.2.

[70] Southwest Center for Biological Diversity v. U.S. Forest Service, 100 F. 3d 1443, 1450 (9th Cir. 1996).

[71] P.L. 104-19, 109 Stat. 194, 240-247.

[72] Rhodes v. Johnson, 153 F. 3d 785 (7th Cir. 1998).

[73] The Report on the Public Comments can be reached via the June 26, 2002 News Release at [http://www.roadless.fs.fed.us]. The Ninth Circuit pointed out that the Attorney General of Montana had asserted that nationally "96% of commenters favored stronger protections." Kootenai Tribe of Idaho v. Veneman, 313 F. 3d 1094 (9th Cir. 2003).

[74] 313 F. 3d 1094 (9th Cir. 2003).

[75] *Id.* at 1104.

[76] *Id.* at 1126.

[77] *Id.* at 1118-1119.

[78] *Id.* at 1120-1121.

[79] Under Rule 24 of the Federal Rules of Civil Procedure, a potential intervenor must meet certain criteria to intervene, either as of right under Rule 24(a) or with the permission of the court under Rule 24(b). Earlier cases, including two in the Ninth Circuit, had held that only the federal government can defend the adequacy of its NEPA compliance. Churchill v. Babbitt, 150 F. 3d 1072, as amended by 158 F. 3d 491 (9th Cir. 1998) held

that the district court in that instance did not err in allowing intervenors under Rule 24(a) to intervene only as to the remedial part of the case. In Portland Audubon Society v. Hodel, 866 F.2d 302 (9th Cir. 1989) the court held that environmental intervenors did not qualify under Rule 24(a) to intervene as of right to defend a NEPA challenge although they evidently were allowed intervenor status on other claims. This latter case referred to an earlier Seventh Circuit case (Wade v. Goldschmidt, 673 F. 2d 182 (7th Cir. 1982)) in which the court denied intervenor status to an applicant because it failed to assert an interest sufficient to warrant intervention as of right under Rule 24(a) in the context of a NEPA challenge, stating: "In a suit such as this, brought to require compliance with federal statues regulating governmental projects, the governmental bodies charged with compliance can be the only defendants." (Wade, at 185.) Furthermore, the court found that "as it should be clear from our discussion of intervention of right," the applicants did not have "a question of law or fact in common" to satisfy the requirement for permissive intervention under Rule 24(b). The dissenting opinion in the Roadless Rule case questioned whether the majority adequately established that the appealing intervenors fit within even permissive intervention under Rule 24(b). The majority held that the district court erred to the extent it permitted intervention under Rule 24(a), but found intervention proper under Rule 24(b). In reaching this conclusion, the court quoted from a leading treatise which seems to postulate generous grounds for allowing intervention. (Kootenai, at 1109.) In a search for "independent jurisdictional grounds"sufficient to support intervention to pursue an appeal abandoned by the other parties, the court looked to the standing of the intervenor applicants. The court determined that the applicants need not show that they independently could have sued the party who prevailed in district court, but need allege only a threat of injury stemming from the order they seek to reverse, an injury which would be redressed if they win on appeal. (*Id.*) The court stated that "intervenors asserted their interests related to the Roadless Rule in moving to intervene" (*Id.* at 1111), but did not clearly set out what those interests were. Possibly, they are the "interest in the use and enjoyment of roadless lands and in the conservation of roadless lands in the national forest lands subject to the roadless Rule" the court mentioned previously. The court also discussed the fact that the district court expressly noted the magnitude of the case and that "the applicants' intervention will contribute to the equitable resolution of this case," to which opinion the appellate court added that the presence of intervenors would "assist the court in its orderly procedures leading to the resolution of this case, which impacted large and varied interests." This approach echoes another Ninth Circuit opinion that had applied a generous approach to intervention saying: "[a] liberal policy in favor of intervention serves both efficient resolution of issues and broadened access to the courts. By allowing parties with a practical interest in the outcome of a particular case to intervene, we often prevent or simplify future litigation involving related issues; at the same time we allow an additional interested party to express its views before the court." (Forest Conservation council v. Forest Service, 66 F. 3d 1489, 1496 n.8 (9th Cir. 1995)(citation omitted)). This same court, at 1497, approached the 'interest' test of Rule 24(b) generously as "primarily a practical guide to disposing of lawsuits by involving as many apparently concerned persons as is compatible with efficiency and due process." The Ninth Circuit denied Idaho's petition for rehearing, and the case was remanded to the district court to reconsider its previous reasoning and injunction in light of the opinion of the appellate court. Several of the other lawsuits challenging the Roadless Rule were stayed pending the decision by the Ninth Circuit.

[80] 68 Fed. Reg. 41865 (July 15, 2003).

[81] 68 Fed. Reg. 41864 (July 15, 2003).

[82] 68 Fed. Reg. 75136 (Dec. 30, 2003).

[83] See [http://www.fs.fed.us/r10/tongass/forest_facts/faqs/roadless.shtml].

[84] Wyoming v. United States Department of Agriculture, 277 F. Supp. 2d 1197 (D. Wyo. (2003).

[85] *Id.* at 1231 - 1232.

[86] *Id.* at 1203.

[87] *Id.* at 1232.

[88] *Id.* at 1239.

[89] Wyoming v. United States Department of Agriculture, 414 F.3d 1207 (10th Cir. 2005).

[90] 69 Fed. Reg. 42,636 (July 16, 2004).

[91] 69 Fed. Reg. 42,648.

[92] 70 Fed. Reg. 1,023.

[93] 36 C.F.R. § 219.27(b)(2004).

[94] *Id.*, at 1,044.

[95] Section 219.12 at 1,059.

[96] Supplemental Response to Public Comments on the 2002 Proposed Planning Rule, at 31.

[97] *Id.*, at 1,044.

[98] *Id.*

[99] 70 Fed. Reg. 14,637.

[100] Although directives purportedly are available at [http://www.fs.fed.us/im/directives/], the actual text of the new directives is unavailable at that site as of April 19, 2005. If a citizen visits the directives site mentioned in the Federal Register [http://www.fs.fed.us/emc/nfma/], the text of the new interim directives is available, but the sequence of the relevant IDs is not evident and the relationship between previous and current IDs is not clear.

[101] 69 Fed. Reg. 42,648 (July 16, 2004).

[102] 70 Fed. Reg. 25,658.

[103] I.D. 1920-2006-1, effective January 16, 2006.

[104] 70 Fed. Reg. 25,654.

[105] See [http://www.roadless.fs.fed.us/].

[106] Don Thompson, *Schwarzenegger Seeks to Maintain Roadless Areas in Forests,* S.F. Gate, July 11, 2006.

[107] See CRS Report RL32436, *Public Participation in the Management of Forest Service and Bureau of Land Management Lands: Overview and Recent Changes*, by Pamela Baldwin.

[108] 70 Fed. Reg. 25,654.

[109] *Id.*

[110] 70 Fed. Reg. 25,656.

[111] 70 Fed. Reg. 25,663.

[112] Biographical information on the thirteen individuals serving as Committee members as of July 6, 2006 can be accessed at [http://www.roadless.fs.fed.us/documents/adv-comm/RACNAC-Bios.pdf].

[113] 70 Fed Reg. 70,580.

[114] States of California et al. v. U.S. Department of Agriculture, et al., 3:05-CV-03508 (Cal. D. August 30, 2005).

[115] Montana and Maine have filed an amicus brief on behalf of plaintiffs; Alaska and Wyoming have filed an amicus brief on behalf of defendants.

[116] *States of California et al.*

[117] Letter to Ann Venemen, U.S. Secretary of Agriculture (August 16, 2004).

[118] In expressing concern over the use of a state-based rule scheme to manage interstate roadless areas, the Governor stated, "The proposal's state-specific rulemaking could result in inconsistent management plans due to conflicting state priorities. Actions on one side of the border will undoubtedly impact and could potentially undermine management strategies on the other side."

[119] Letter from Theodore Kulongoski, Governor of Oregon, to Mike Johanns, U.S. Secretary of Agriculture (October 14, 2005). On October 27, 2005, the Department denied the petition for an expedited process scheme, expressing doubt that the Governor's proposed rule would substantially decrease states' burdens. Letter from Mark Rey, Under Secretary of Natural Resources and Environment, U.S. Department of Agriculture, to Theodore Kulongoski, Governor of Oregon.

[120] Letter to U.S. Secretary of Agriculture Ann Venemen, Sept 8, 2004.

[121] Letter to U.S. Secretary of Agriculture Ann Venemen, July 30, 2004. The Forest Service has provided financial assistance to several states for costs incurred relating to petition development. On December 20, 2005, the FS announced a $150,000 grant to the State of Idaho, and on June 28, 2006, it announced a $200,000 grant to the State of Arizona. The grants were in response to formal requests by the states for financial support for the petitioning process. U.S. Department of Agriculture Press Releases Nos. 0560.05 and 0225.06.

[122] USDA News Release No. 0212.06.

[123] Letter from Mark Rey, Under Secretary of Natural Resources and Environment, U.S. Department of Agriculture, to Timothy Kaine, Governor of Virginia (June 21, 2006); Letter from Rey to Michael Easley, Governor of North Carolina (June 21, 2006); Letter from Rey to Mark Sanford, Governor of South Carolina (June 21, 2006).

[124] The Save America's Forests Act, S. 1897, 109th Cong. (2005). The last major Congressional action on this bill was taken on October 19, 2005, when it was referred to the Senate Committee on Energy and Natural Resources.

[125] Roadless Areas Conservation Act of 2006, S. 2364, 109th Cong. (2006). The last major Congressional action on this was taken on March 3, 2006, when it was referred to the Senate Committee on Energy and Natural Resources.

In: Environmental Research Advances
Editor: Peter A. Clarkson, pp. 77-94

ISBN: 978-1-60021-762-3
© 2007 Nova Science Publishers, Inc.

Chapter 3

WETLANDS:
AN OVERVIEW OF ISSUES*

Jeffrey A. Zinn and Claudia Copeland

ABSTRACT

The 109th Congress, like earlier ones, continues to consider numerous policy topics that involve wetlands. Of interest are statements by the Bush Administration shortly after the 2004 election that restoration of 3 million wetland acres would be a priority. Wetland policies continue to attract congressional interest, and in recent months, that interest has become focused on the role that restored wetlands could play in protecting New Orleans, and coastal Louisiana more generally, from hurricanes.

In the 108th Congress and so far in the 109th Congress, no major wetlands legislation has been enacted. Earlier Congresses had reauthorized and amended many wetland programs and examined controversies such as applying federal regulations on private lands, wetland loss rates, implementation of farm bill provisions, and proposed changes to the federal permit program.

Congress also has been involved at the program level, responding to legal decisions and administrative actions. Examples include implementation of Corps of Engineers changes to the nationwide permit program; redefining key wetlands permit regulatory terms in revised rules issued in 2002; and a 2001 Supreme Court ruling (called the *SWANCC* case) that narrowed federal regulatory jurisdiction over certain isolated wetlands. Hearings on many of these topics were held, and some legislation was introduced. Legislation to reverse the *SWANCC* ruling has been introduced (H.R. 1356, the Clean Water Authority Restoration Act), as has a bill to narrow the government's regulatory jurisdiction (H.R. 2658, the Federal Wetlands Jurisdiction Act). A June 2006 Supreme Court ruling in two related cases could accelerate congressional attention to these issues.

Wetland protection efforts engender intense controversy over issues of science and policy. Controversial topics include the rate and pattern of loss, whether all wetlands should be protected in a single fashion, the ways in which federal laws currently protect them, and the fact that 75% of remaining U.S. wetlands are located on private lands.

---

* Excerpted from CRS Report RL33483, dated June 20, 2006.

One reason for these controversies is that wetlands occur in a wide variety of physical forms, and the numerous values they provide, such as wildlife habitat, also vary widely. In addition, the total wetland acreage in the lower 48 states is estimated to have declined from more than 220 million acres three centuries ago to 107.7 million acres in 2004. The long-standing national policy goal of no-net-loss has been reached, according to the Fish and Wildlife Service, as the rate of loss has been more than offset by net gains through expanded restoration efforts authorized in multiple laws. Many protection advocates view these laws as inadequate or uncoordinated. Others, who advocate the rights of property owners and development interests, characterize them as too intrusive. Numerous state and local wetland programs add to the complexity of the protection effort.

MOST RECENT DEVELOPMENTS

The U.S. Fish and Wildlife Service released its most recent periodic survey of changes in wetland acreage in March 2006. Covering 1998 to 2004, it concluded that during this time period there was a small net gain in overall wetland acres for the first time in this survey. Others caution, however, that much of this gain was in ponds, rather than natural wetlands.

Hurricanes Katrina and Rita caused widespread alteration and destruction of wetlands along the central Gulf Coast. The net effect will likely be major permanent losses, especially along the coast. These losses will be partially offset as some destruction will be temporary and other new wetlands are created. The extent of change and loss is being documented by federal agencies and others [1]. Congress is considering numerous alternative legislative proposals that would fund wetland restoration projects and activities to help lessen the impact of future hurricanes. The 109th Congress had been considering a set of proposals to restore coastal wetlands in Louisiana, and both the proposals and the funding level have been expanded as a result of these hurricanes.

In the 109th Congress, about five dozen bills with wetland provisions have been introduced; about two dozen of these address wetland loss and restoration along the central Gulf Coast. The remainder address topics that attracted attention in earlier Congresses, but were not acted on, including legislation to reverse a controversial 2001 Supreme Court ruling concerning isolated wetlands, the *SWANCC* case (H.R. 1356, the Clean Water Authority Restoration Act); legislation to narrow the government's regulatory jurisdiction (H.R. 2658, the Federal Wetlands Jurisdiction Act); other large-scale restoration efforts involving wetlands (the Everglades, for example); and appropriations for wetland programs. Concerning the *SWANCC* case, critics say that guidance issued by the Environmental Protection Agency (EPA) in 2003 interpreting the case for field staff goes beyond what the Supreme Court's decision required and has allowed many streams and wetlands to be unprotected from development. On May 18, 2006, the House adopted an amendment to H.R. 5386 to prohibit EPA from spending funds to implement the controversial guidance.

Federal courts have had a key role in interpreting and clarifying the limits of federal jurisdiction to regulate activities that affect wetlands, especially since the *SWANCC* decision. On February 21, the Supreme Court heard arguments in two cases brought by landowners (*Rapanos v. United States*; *Carabell v. U.S. Army Corps of Engineers*) seeking to narrow the scope of the Clean Water Act (CWA) permit program as it applies to development of wetlands. The Court's ruling was issued on June 19. In a 5-4 decision, a plurality of the Court held that the lower court had applied an incorrect standard to determine whether the wetlands at issue are covered by the CWA. Justice Kennedy joined this plurality to vacate the lower court decisions and remand the cases for further consideration, but he took different positions on most of the substantive issues raised by the cases, as did four dissenting justices, leading to uncertainty about interpretation and implications of the ruling.

BACKGROUND AND ANALYSIS

Wetlands, with a variety of physical characteristics, are found throughout the country. They are known in different regions as swamps, marshes, fens, potholes, playa lakes, or bogs. Although these places can differ greatly, they all have distinctive plant and animal assemblages because of the wetness of the soil. Some wetland areas may be continuously inundated by water, while other areas may not be flooded at all. In coastal areas, flooding may occur on a daily basis as tides rise and fall.

Functional values, both ecological and economic, at each wetland depend on its location, size, and relationship to adjacent land and water areas. Many of these values have been recognized only recently. Historically, many federal programs encouraged wetlands to be drained or altered because they were seen as having little value as wetlands. Wetland values can include:

- habitat for aquatic birds and other animals and plants, including numerous threatened and endangered species; production of fish and shellfish;
- water storage, including mitigating the effects of floods and droughts;
- water purification; ! recreation;
- timber production;
- food production;
- education and research; and
- open space and aesthetic values.

Usually wetlands provide some combination of these values; no single wetland in most instances provides all these values. The composite value typically declines when wetlands are altered. In addition, the effects of alteration often extend well beyond the immediate area because wetlands are usually part of a larger water system. For example, conversion of wetlands to urban uses has increased flood damages; this value is receiving considerable attention as natural disaster costs have mounted through the 1990s.

Federal laws that affect wetlands have changed since the mid-1980s as the values of wetlands have been recognized in different ways in numerous national policies. Previously, some laws encouraged destruction of wetland areas, such as selected provisions in the federal tax code, public works legislation, and farm programs. Federal laws now either encourage wetland protection, or prohibit or do not support their destruction. These laws, however, do not add up to a fully consistent or comprehensive national approach. The central federal regulatory program, Section 404 of the Clean Water Act, requires permits for the discharge of dredged or fill materials into many but not all wetland areas. However, other activities that may adversely affect wetlands do not require permits, and some places that scientists define as wetlands are exempt from this permit program because of physical characteristics. An agricultural program, swampbuster, is a disincentive program that indirectly protects wetlands by making farmers who drain wetlands ineligible for federal farm program benefits; those who do not receive these benefits have no reason to observe the requirements of this program. Several land acquisition and other incentive programs complete the current federal protection effort.

Although numerous wetland protection bills have been introduced in recent Congresses, the most significant new wetlands legislation to be enacted has been in the two most recent farm bills, in 1996 and 2002. During this period, Congress also reauthorized several wetlands programs, mostly setting higher appropriations ceilings, without making significant shifts in policy. President Bush endorsed wetland protection in signing the farm bill and the North American Wetlands Conservation Act reauthorization in 2002. The Bush Administration has issued guidance on mitigation policies and regulatory program jurisdiction; the latter has raised controversy with some groups (see discussion below).

In 2002, the Bush Administration endorsed the concept of "no-net-loss" of wetlands — a goal declared by President George H. W. Bush in 1988 and also embraced by President Clinton to balance wetlands losses and gains in the short term and achieve net gains in the long term. On Earth Day 2004, the President announced a new national goal, moving beyond no-net-loss, of achieving an overall increase of wetlands [2]. The goal is to create, improve, and protect at least three million wetland acres over the next five years in order to increase overall wetland acres and quality. (By comparison, the Clinton Administration in 1998 announced policies intended to achieve overall wetland increases of 200,000 acres per year by 2005.) To meet the new goal, President Bush urged Congress to pass his FY2005 budget request for conservation programs, and in which he focused on two wetlands programs, the Wetlands Reserve Program (WRP) and the North American Wetlands Conservation Act Grants Program (NAWCP). The FY2005 budget request for these two programs, $349 million, was 10% more than FY2004 levels. (However, Congress disagreed, providing level funding for the NAWCP and an 18% reduction for the WRP.) The President's strategy also calls for better tracking of wetland programs and enhanced local and private sector collaboration.

In April 2005, the Administration issued a report saying that about 832,000 acres of wetlands have been created, protected, or improved in the past year as part of the President's program, and another 1.6 million acres are expected to be added by the end of FY2006 [3] Environmental groups criticized the report as presenting an incomplete picture, because it fails to mention wetlands lost to agriculture and development.

Congress has provided a forum in numerous hearings where conflicting interests in wetland issues have been debated. Broadly speaking, the conflicts are between:

- Environmental interests and wetland protection advocates who have been pressing for greater wetlands protection as multiple values have been more widely recognized, by improving coordination and consistency among agencies and levels of governments, and strengthened programs; and
- Others, including landowners, farmers, and small businessmen, who counter that protection efforts have gone too far, and that privately owned wet areas that provide few wetland values have been aggressively protected. They have been especially critical of the U.S. Army Corps of Engineers (Corps) and the U.S. Environmental

Protection Agency (EPA), asserting that they administer the Section 404 program in an overzealous and inflexible manner.

Wetland issues revolve around disparate scientific and programmatic questions, and conflicting views of the role of government where private property is involved. Scientific questions include how to define wetlands, the current rate and pattern of wetland declines and losses, and the importance of these physical changes. Federal program issues include the administration of programs to protect, restore, or mitigate wetland resources (especially the Clean Water Act Section 404 program); relationships between agriculture and wetlands; whether all wetlands should be treated the same in federal programs and which wetlands should be subject to regulation; federal funding of wetland programs; and is whether protecting wetlands by acres is a good proxy for protecting wetlands based on the functions they perform and the values they provide. In addition, private property questions are raised because almost three-quarters of the remaining wetlands are located on private lands, and some property owners believe they should be compensated when federal programs limit how they can use their land, and thereby diminish its value.

What Is a Wetland?

There is general agreement that scientists can determine the presence of a wetland by a combination of soils, plants, and hydrology. The only definition of wetlands in law, in the swampbuster provisions of farm legislation (P.L. 99-198) and reproduced in the Emergency Wetlands Resources Act of 1986 (P.L. 99-645), lists those three components but does not include more specific criteria, such as what conditions must be present and for how long. Controversies are exacerbated when many sites that have those three components and are identified as wetlands by experts, either may have wetland characteristics only some portion of the time, or may not look like what many people visualize as wetlands.

Wetlands subject to federal regulation are a large subset of all places that the scientific community would call a wetland. These regulated wetlands, under the Section 404 program discussed below, are currently identified using technical criteria in a wetland delineation manual issued by the Corps in 1987. It was prepared jointly and is used by all federal agencies to carry out their responsibilities under this program (the Corps, EPA, Fish and Wildlife Service (FWS), and the National Marine Fisheries Service (NMFS)). The manual provides guidance and field-level consistency among the agencies that have roles in wetland regulatory protection. (A second and slightly different manual, agreed to by the Corps and the Natural Resources Conservation Service, is used for delineating wetlands on agricultural lands.) While the agencies try to improve the objectivity and consistency of wetland identification and delineation, judgement continues to play a role and can lead to site-specific controversies. Cases discussed below (see "Judicial Proceedings Involving Section 404") are efforts to exclude wetlands in certain physical settings or certain activities affecting them from the regulatory program.

How Fast Are Wetlands Disappearing, and How Many Acres Are Left?

The U.S. Fish and Wildlife Service periodically surveys national net trends in wetland acreage using the National Wetlands Inventory (NWI). It has estimated that when European settlers first arrived, wetland acreage in the area that would become the 48 states was more than 220 million acres, or about 5% of the total land area. By 2004, total wetland acreage was estimated to be 107.7 million acres, according to data it presented in its most recent survey [4]. Data compiled by the NRCS and the FWS in separate surveys and using different methodologies have identified similar trends. Both show that the annual net loss rate dropped from almost 500,000 acres annually nearly three decades ago to slight net annual gains in recent years. The FWS survey estimated the average annual gain between 1998 and 2004 was 32,000 acres, primarily associated with the expansion of shallow ponds, while NRCS (using its Natural Resources Inventory (NRI) of privately-owned lands) estimated that there was an average annual gain of 26,000 acres between 1997 and 2002. NRCS cautioned against making precise claims of net increases because of statistical uncertainties. Some environmentalists caution that the increases identified in the latest FWS data are tied to a proliferation of small ponds rather than natural wetlands.

Numerous shifts in federal policies since 1985 (and changes in economic conditions as well) strongly influence wetland loss patterns, but the composite effects remain unmeasured beyond these raw numbers. There usually is a large time lag from the announcement and implementation of changes in policy to collection and release of data that measure how these changes affect loss rates. Also, it is often very difficult to distinguish the role that policy changes play from other factors, such as agricultural markets, development pressures, and land markets.

Further, these data only measure acres. They do not provide any insights into changes in the quality of remaining wetlands as measured by the values they provide, which is often determined by where a wetland is located in a watershed, surrounding land uses, etc. Nevertheless, in his Earth Day 2004 wetlands announcement (discussed above), President Bush said that as the nation is nearing the goal of no-net-loss, it is appropriate to move towards policies that will result in a net increase of wetland acres and quality.

The Clean Water Act Section 404 Program

The principal federal program that provides regulatory protection for wetlands is found in Section 404 of the Clean Water Act (CWA). Its intent is to protect water and adjacent wetland areas from adverse environmental effects due to discharges of dredged or fill material. Established in 1972, Section 404 requires landowners or developers to obtain permits from the Corps of Engineers to carry out activities involving disposal of dredged or fill materials into waters of the United States, including wetlands.

The Corps has long had regulatory jurisdiction over dredging and filling, starting with the River and Harbor Act of 1899. The Corps and EPA share responsibility for administering the Section 404 program. Other federal agencies, including NRCS, FWS, and NMFS, also have roles in this process. In the 1970s, legal decisions in key cases led the Corps to revise this program to incorporate broad jurisdictional definitions in terms of both regulated waters and adjacent wetlands. Section 404 was last amended in 1977.

This judicial/regulatory/administrative evolution of the Section 404 program has generally pleased those who view it as a critical tool in wetland protection, but dismayed others who would prefer more limited Corps jurisdiction or who see the expanded regulatory program as intruding on private land-use decisions and treating wetlands of widely varying value similarly. Underlying this debate is the more general question of whether Section 404 is the best approach to federal wetland protection.

Some wetland protection advocates have proposed that it be replaced or greatly altered. First, they point out that it governs only the discharge of dredged or fill material, while not regulating other acts that drain, flood, or otherwise reduce functional values. Second, because of exemptions provided in 1977 amendments to Section 404, major categories of activities are not required to obtain permits. These include normal, ongoing farming, ranching, and silvicultural (forestry) activities. Further, permits generally are not required for activities which drain wetlands (only for those that fill wetlands), which excludes a large number of actions with potential to alter wetlands. Third, in the view of protection advocates, the multiple values that wetlands can provide (e.g., fish and wildlife habitat, flood control) are not effectively recognized through a statutory approach based principally on water quality, despite the broad objectives of the Clean Water Act.

The Permitting Process. The Corps' regulatory process involves both general permits for actions by private landowners that are similar in nature and will likely have a minor effect on wetlands and individual permits for more significant actions. According to the Corps, it evaluates more than 85,000 permit requests annually. Of those, more than 90% are authorized under a general permit, which can apply regionally or nationwide, and is essentially a permit by rule, meaning the proposed activity is presumed to have a minor impact. Most do not require pre-notification or prior approval. About 9% are required to go through the more detailed evaluation for a standard individual permit, which may involve complex proposals or sensitive environmental issues and can take 180 days or longer for a decision. Less than 0.3% of permits are denied; most other individual permits are modified or conditioned before issuance. About 5% of applications are withdrawn prior to a permit decision. In FY2003 (the most recent year for which data are available), Corps-issued permits authorized activities having a total of 21,330 acres of wetland impact, while those permits required that 43,379 acres of wetlands be restored, created, or enhanced as mitigation for the authorized losses[5].

Regulatory procedures on individual permits allow for interagency review and comment, a coordination process that can generate delays and an uncertain outcome, especially for environmentally controversial projects. EPA is the only federal agency having veto power over a proposed Corps permit; EPA has used its veto authority 11 times in the 30-plus years since the program began. Critics have charged that implied threats of delay by the FWS and others practically amount to the same thing. Reforms during the Reagan, earlier Bush, and Clinton Administrations streamlined certain of these procedures, with the intent of speeding up and clarifying the Corps' full regulatory program, but concerns continue over both process and program goals.

Controversy also surrounded revised regulations issued by EPA and the Corps in May 2002, which redefine two key terms in the 404 program: "fill material" and "discharge of fill material." The agencies said that the revisions were intended to clarify certain confusion in their joint administration of the program due to previous differences in how the two agencies defined those terms. However, environmental groups contended that the changes allow for less restrictive and inadequate regulation of certain disposal activities, including disposal of coal mining waste, which could be harmful to aquatic life in streams. The Senate Environment and Public Works Committee held a hearing in June 2002 to review these issues, and legislation to reverse the agencies' action was introduced, but no further action occurred [6]. That legislation was re-introduced in the 108th Congress, and again in the 109th Congress (H.R. 2719).

Nationwide Permits. Nationwide permits are a key means by which the Corps minimizes the burden of its regulatory program. A nationwide permit is a form of general permit which authorizes a category of activities throughout the nation and is valid only if the conditions applicable to the permit are met. These general permits authorize activities that are similar in nature and are judged to cause only minimal adverse effect on the environment. General permits minimize the burden of the Corps' regulatory program by authorizing landowners to proceed without having to obtain individual permits in advance.

The current program has few strong supporters, for differing reasons. Developers say that it is too complex and burdened with arbitrary restrictions. Environmentalists say that it does not adequately protect aquatic resources. At issue is whether the program has become so complex and expansive that it cannot either protect aquatic resources or provide for a fair regulatory system, which are its dual objectives.

Nationwide permits are issued for periods of no longer than five years and thereafter must be reissued by the Corps. The most recent reissuance, in January 2002, included some changes, including relaxation of certain permit conditions, intended by the Corps to add flexibility. Reactions to the permits were mixed: environmental advocates contend that the re-issued permits are not adequately protective of water quality and will result in a net loss of wetland acres, while developer groups argue that the overall program continues to focus on arbitrary regulatory thresholds that result in undue burden on developers and the Corps [7]. In July 2005, a federal court of appeals panel held that the Corps' issuance of the 2000 and 2002 nationwide permits constitutes final agency action, thus permitting a challenge to the permits that had been brought by developers to proceed.

Citizen groups have filed lawsuits seeking to halt the Corps' use of one of its nationwide permits, NWP 21, to authorize a type of coal mining practice called mountaintop mining. In 2004, a federal district court in West Virginia ruled that NWP 21 violates the CWA by authorizing activities that have more than minimal adverse environmental effects. The district court's ruling was overturned on appeal, but a request for rehearing is pending. Another lawsuit challenging the applicability of nationwide permits to mountaintop mining in Kentucky also has been filed [8].

Section 404 authorizes states to assume many of the permitting responsibilities. Two states, Michigan (in 1984) and New Jersey (in 1992), have done this. Others have cited the complex process of assumption, the anticipated cost of running a program, and the continued involvement of federal agencies because of statutory limits on waters that states could regulate as reasons for not joining these two states. Efforts continue toward encouraging more states to assume program responsibility.

Judicial Proceedings Involving Section 404: *SWANCC.* The Section 404 program has been the focus of numerous lawsuits, most of which have sought to narrow the geographic scope of the regulatory program. In that context, an issue of long-standing controversy is whether isolated waters are properly within the jurisdiction of Section 404. Isolated waters (those that lack a permanent surface outlet to downstream waters) which are not physically adjacent to navigable surface waters often appear to provide few of the values for which wetlands are protected, even if they meet the technical definition of a wetland. In January 2001, the Supreme Court ruled on the question of whether the CWA provides the Corps and EPA with authority over isolated waters and wetlands. The Court's 5-4 ruling in *Solid Waste Agency of Northern Cook County (SWANCC) v. U.S. Army Corps of Engineers* (531 U.S. 159) held that the denial of a Section 404 permit for disposal on isolated wetlands solely on the basis that migratory birds use the site exceeds the authority provided in the act. The full extent of retraction of the regulatory program resulting from this decision remains unclear, even five years after the ruling. Environmentalists believe that the Court misinterpreted congressional intent on the matter, while industry and landowner groups welcomed the ruling [9].

Policy implications of how much the decision restricts federal regulation depend on how broadly or narrowly the opinion is applied, and since the 2001 Court decision, other federal courts have issued a number of rulings that have reached varying conclusions. Some federal courts have interpreted *SWANCC* narrowly, thus limiting its effect on current permit rules, while a few read the decision more broadly. However, in April 2004, the Court declined to review three cases that support a narrow interpretation of *SWANCC*. Environmentalists were pleased that the Court rejected the petitions, but attorneys for industry and developers say that the courts will remain the primary battleground for CWA jurisdiction questions, so long as neither the Administration nor Congress takes steps to define jurisdiction.

The government's current view on the key question of the scope of CWA jurisdiction in light of SWANCC and other court rulings came in a legal memorandum issued jointly by EPA and the Corps on January 15, 2003 [10]. It provides a legal interpretation essentially based on a narrow reading of the Court's decision, thus allowing federal regulation of some isolated waters to continue (in cases where factors other than the presence of migratory birds may exist, thus allowing for assertion of federal jurisdiction), but it calls for more review by higher levels in the agencies in such cases. Administration press releases say that the guidance demonstrates the government's commitment to "no-net-loss" wetlands policy. However, it was apparent that the issues remained under discussion, because at the same time, the Administration issued an advance notice of proposed rulemaking (ANPRM) seeking comment on how to define waters that are under jurisdiction of the regulatory program. The ANPRM did not actually propose rule changes, but it indicated possible ways that Clean Water Act rules might be modified to further limit federal jurisdiction, building on *SWANCC* and some subsequent legal decisions. The government received more than 133,000 comments on the ANPRM, most of them negative, according to EPA and the Corps. Environmentalists and many states opposed changing any rules, saying that the law and previous court rulings call for the broadest possible interpretation of the Clean Water Act (and narrow interpretation of *SWANCC*), but developers sought changes to clarify interpretation of the *SWANCC* ruling.

In December 2003, EPA and the Corps announced that the Administration will not pursue rule changes concerning federal regulatory jurisdiction over isolated wetlands. The EPA Administrator said that the Administration wanted to avoid a contentious and lengthy rulemaking debate over the issue. Environmentalists and state representatives expressed relief at the announcement. Interest groups on all sides have been critical of confusion in implementing the 2003 guidance, which constitutes the main tool for interpreting the reach of the *SWANCC* decision. Environmentalists remain concerned about diminished protection resulting from the guidance, while developers said that without a new rule, confusing and contradictory interpretations of wetland rules likely will continue. In that vein, a Government

Accountability Office (GAO) report concluded that Corps districts differ in how they interpret and apply federal rules when determining which waters and wetlands are subject to federal jurisdiction, documenting enough differences that the Corps has begun a comprehensive survey of its district office practices to help promote greater consistency [11] Concerns over inconsistent or confusing regulation of wetlands have also drawn congressional interest [12].

In response to continuing controversies about the 2003 guidance, on May 18, 2006, the House adopted an amendment to a bill providing FY2007 appropriations for EPA (H.R. 5386). The amendment (passed by a 222-198 vote) would bar EPA from spending funds to implement the 2003 policy guidance. Senate action is pending. Supporters of the amendment said that the guidance goes beyond what the Supreme Court required in *SWANCC,* has allowed many streams and wetlands to be unprotected from development, and has been more confusing than helpful. Opponents of the amendment predicted that it would make EPA's and the Corps' regulatory job more difficult than it already is.

While the issue of how regulatory protection of wetlands is affected by the *SWANCC* decision and subsequent developments continues to evolve, the remaining responsibility to protect affected wetlands falls on states and localities. Whether states will act to fill in the gap left by removal of some federal jurisdiction is likely to be constrained by budgetary and political pressures, but a few states (Wisconsin and Ohio, for example) have passed new laws or amended regulations to do so. In comments on the ANPRM, many states said that they do not have authority or financial resources to protect their wetlands, in the absence of federal involvement.

Federal courts continue to have a key role in interpreting and clarifying the *SWANCC* decision. On February 21, 2006, the Supreme Court heard arguments in two cases brought by landowners (*Rapanos v. United States*; *Carabell v. U.S. Army Corps of Engineers*) seeking to narrow the scope of the CWA permit program as it applies to development of wetlands. The issue in both cases had to do with the reach of the CWA to cover "waters" that were not navigable waters, in the traditional sense, but were connected somehow to navigable waters or "adjacent" to those waters. (The act requires a federal permit to discharge dredged or fill materials into "navigable waters.") Many legal and other observers hoped that the Court's ruling in these cases would bring greater clarity about the scope of federal regulatory jurisdiction [13].

The Court's ruling was issued on June 19 (*Rapanos et ux., et al., v. United States*, Nos. 04-1034 and 04-1384, 547 U.S. ___). In a 5-4 decision, a plurality of the Court, led by Justice Scalia, held that the lower court had applied an incorrect standard to determine whether the wetlands at issue are covered by the CWA. Justice Kennedy joined this plurality to vacate the lower court decisions and remand the cases for further consideration, but he took different positions on most of the substantive issues raised by the cases, as did four other dissenting justices. Early judgments by legal observers suggest that the implications of the ruling (both short-term and long-term) are far from clear. Because the several opinions written by the justices did not draw a clear line regarding what wetlands and other waters are subject to federal jurisdiction, one likely result is more case-by-case determinations and continuing litigation. There also could be renewed pressure on the Corps and EPA to clarify the issues through an administrative rulemaking.

Legislation to reverse the *SWANCC* decision has been introduced in the 109th Congress (H.R. 1356, the Clean Water Authority Restoration Act of 2005); identical legislation was introduced in the 108th Congress (H.R. 962, S. 473). It would provide a broad statutory definition of "waters of the United States;" clarify that the CWA is intended to protect U.S. waters from pollution, not just maintain their navigability; and include a set of findings to assert constitutional authority over waters and wetlands. Other legislation to restrict regulatory jurisdiction also has been introduced in the 109th Congress (H.R. 2658, the Federal Wetlands Jurisdiction Act of 2005). It would narrow the statutory definition of "navigable waters" and define certain isolated wetlands and other areas as not being subject to federal regulatory jurisdiction. It also would give the Corps sole authority to determine §404 jurisdiction, for permitting purposes. Similar legislation also was introduced in the 108th Congress (H.R. 4843). For now, it is unclear whether the more recent decision in the *Rapanos* and *Carabell* cases will accelerate congressional interest in these or other proposals to address uncertainties about federal jurisdiction over wetlands and other waters.

Should All Wetlands Be Treated Equally? Under the Section 404 program, there is a perception that all jurisdictional wetlands are treated equally, regardless of size, functions, or values. This has led critics to focus on situations where a wetland has little apparent value, but the landowner's proposal is not approved or the landowner is penalized for altering a wetland without a federal permit. Critics believe that one possible solution may be to have a tiered approach for regulating wetlands. Several legislative proposals introduced in recent Congresses would establish multiple tiers (typically three) — from highly valuable wetlands that should receive the greatest protection to the least valuable wetlands where alterations might usually be allowed. Some states (New York, for example) use such an approach for state-regulated wetlands. The Corps and EPA issued guidance to field staff emphasizing the flexibility that currently exists in the Section 404 program to apply less vigorous permit review to small projects with minor environmental impacts.

Three questions arise: (1) What are the implications of implementing a classification program? (2) How clearly can a line separating each wetland category be defined? (3) Are there regions where wetlands should be treated differently? Regarding classification, even most wetland protection advocates acknowledge that there are some situations where a wetland designation with total protection is not appropriate. But they fear that classification for different degrees of protection could be a first step toward a major erosion in overall wetland protection. Also, these advocates would probably like to see almost all wetlands presumed to be in the highest protection category unless experts can prove an area should receive a lesser level of protection, while critics who view protection efforts as excessive, would seek the reverse.

Locating the boundary line of a wetland can be controversial when the line encompasses areas that do not meet the image held by many. Controversy would likely grow if a tiered approach required that lines segment wetland areas. On the other hand, a consistent application of an agreed-on definition may lead to fewer disputes and result in more timely decisions.

Some states have far more wetlands than others. Different treatment has been proposed for Alaska because about one-third of the state is designated as wetlands, yet a very small portion has been converted. Legislative proposals have been made to exempt that state from the Section 404 program until 1% of its wetlands have been lost. Some types of wetlands are already treated differently. For example, playas and prairie potholes have somewhat different definitions under swampbuster (discussed below), and the effect is to increase the number of acres that are considered as wetlands. This differential treatment contributes to questions about federal regulatory consistency on private property.

Agriculture and Wetlands

National surveys almost two decades ago indicated that agricultural activities had been responsible for about 80% of wetland loss in the preceding decades, making this topic a focus for policymakers. Congress responded by creating programs in farm legislation starting in 1985 that use disincentives and incentives to encourage landowners to protect and restore wetlands. Swampbuster and the Wetlands Reserve Program are the two largest efforts, but others such as the Conservation Reserve Program's Farmed Wetlands Option and Conservation Reserve Enhancement Program are also being used to protect wetlands. The most recent wetland loss survey conducted by the Natural Resources Conservation Service (NRCS) (comparing data from 1997 and 2002) indicates that there is a small annual increase, for the first time since these data have been collected, of 26,000 acres [14]. However, the agency warns that statistical uncertainties preclude concluding with certainty that gain is actually occurring.

Swampbuster. Swampbuster, enacted in 1985, uses disincentives rather than regulations to protect wetlands on agricultural lands. It remains controversial with farmers concerned about redefining an appropriate federal role in wetland protection on agricultural lands, and with wetland protection advocates concerned about inadequate enforcement. Since 1995, the NRCS has made wetland determinations only in response to requests because of uncertainty over whether changes in regulation or law would modify boundaries that have already been delineated. NRCS has estimated that more than 2.6 million wetland determinations have been made and that more than 4 million may eventually be required.

Swampbuster was amended in the 1996 farm bill (P.L. 104-127) and the 2002 farm bill (P.L. 107-171). Amendments in 1996 granted producers greater flexibility by making changes such as: exempting swampbuster penalties when wetlands are voluntarily restored; providing that prior converted wetlands are not to be considered "abandoned" if they remain in agricultural use; and granting good-faith exemptions. They also encourage mitigation, establish a mitigation banking pilot program, and repeal required consultation with the U.S. Fish and Wildlife Service. The 2002 farm bill made just a single amendment that has not affected either the acres that are protected or the characteristics of the protection effort.

Other Agricultural Wetlands Programs. Under the Wetland Reserve Program (WRP), enacted in 1990, landowners receive payments for placing easements on farmed wetlands. All easements were permanent until provisions in the 1996 farm bill, requiring temporary easements and multi-year agreements as well, were implemented. The 2002 farm bill reauthorized the program through FY2007 and raised the enrollment cap to 2,275,000 acres, with 250,000 acres to be enrolled annually. In addition, in June 2004, NRCS announced a new enhancement program on the lower Missouri River in Nebraska to enroll almost 19,000 acres at a cost of $26 million, working with several public and private partners.

Through FY2005, 9,226 projects had enrolled 1.744 million acres, and easements have been perfected on 1.37 million of those acres. A majority of the easements are in three states: Louisiana, Mississippi, and Arkansas. Most of the land is enrolled under permanent easements, while only about 10% is enrolled under 10-year restoration agreements, according to data supplied by NRCS in support of its FY2007 budget request. Prior to the 2002 farm bill, farmer interest had exceeded available funding, which may help to explain why Congress raised the enrollment ceiling in that legislation.

The 2002 farm bill also expanded the 500,000-acre Farmable Wetlands Pilot Program within the Conservation Reserve Program (CRP) to a 1-million-acre program available nationwide. Only wetland areas that are smaller than 10 acres and are not adjacent to larger streams and rivers are eligible. This program may become more important to overall protection efforts in the wake of the *SWANCC* decision, discussed above, which limited the reach of the Section404 permit program so that it does not apply to many small wetlands that are isolated from navigable waterways. Through April 2006, more than 148,600 acres had been enrolled in this pilot program through more than 9,500 contracts.

On August 4, 2004, the Administration announced a new Wetland Restoration Initiative to allow enrollment of up to 250,000 acres of large wetland complexes and playa lakes located outside the 100-year floodplain in the CRP after October 1, 2004. The Administration estimates implementation of this initiative will cost $200 million. Participants receive incentive payments to help pay for restoring the hydrology of the site, as well as rental payments and cost sharing assistance to install eligible conservation practices.

Several other large agriculture conservation programs, including the Environmental Quality Incentives Program, the Farmland Protection Program, and the Wildlife Habitat Incentive Program, were also amended in the 2002 farm bill in ways that may have incidental protection benefits for wetlands, because of much higher funding levels and because of program changes. Finally, some new programs could less directly help protect wetlands, including the Conservation Security Program, which would provide payments to install and maintain practices on working agricultural lands; a Surface and Groundwater Conservation Program (funded through the Environmental Quality Incentive Program); a new program to retire wetlands that are part of a cranberry operation, and several other programs to better manage water resources [15].

Agricultural Wetlands and the Section 404 Program. The Section 404 program, described above, applies to qualified wetlands in all locations, including agricultural lands. But the Corps and EPA exempt "prior converted lands" (wetlands modified for agricultural purposes before 1985) from Section 404 permit requirements under a memorandum of agreement (MOA), and since 1977 the Clean Water Act has exempted "normal farming activities." The January 2001 Supreme Court *SWANCC* decision, also discussed above, apparently will exempt certain isolated wetlands from Corps jurisdiction; NRCS estimated that about 8 million acres in agricultural locations might be exempted by this decision. In December 2002, the Supreme Court affirmed a lower court decision, without comment, that deep ripping to prepare wetland soils for planting was more than a "normal farming activity" and therefore subject to Section 404 requirements.

While these exemptions and the MOA have displeased some protection advocates, they have probably dampened some of the criticism from farming interests over federal regulation of private lands. On the other hand, how NRCS responds to the *SWANCC* decision on isolated wetlands could cause that criticism to rise. The Corps and NRCS have been unsuccessful in revising the MOA since 1996 despite a decade of negotiation, although they signed a very general partnership agreement on July 7, 2005. Some of the wetlands that fall outside Section 404 requirements as a result of the *SWANCC* decision can now be protected if landowners decide to enroll them into the revised farmable wetlands program or under other new initiatives, described above.

Private Property Rights and Landowner Compensation

An estimated 74% of all remaining wetlands in the coterminous states are on private lands. Questions of federal regulation of private property stem from the argument that land owners should be compensated when a "taking" occurs and alternative uses are prohibited or restrictions on use are imposed to protect wetland values. The U.S. Constitution provides that property owners shall be compensated if private property is "taken" by government action. The courts generally have found that compensation is not required unless all reasonable uses are precluded. Many individuals or companies purchase land with the expectation that they can alter it. If that ability is denied, they contend, then the land is greatly reduced in value. Many argue that a taking should be recognized when a site is designated as a wetland. In 2002, the Supreme Court held that a Rhode Island man, who had acquired property after the state enacted wetlands regulation affecting the parcel, is not automatically prevented from bringing an action to recover compensation from the state. Instead, the court ruled that the property retained some economic use after the state's action. *(Palazzolo v. Rhode Island*, 533 U.S. 606, 2002).

Congress has explored these wetlands property rights issues. An example is an October 2001 hearing by the House Transportation and Infrastructure Committee, Subcommittee on Water Resources and the Environment [16]. Recent Congresses have considered, but did not enact, property rights protection proposals. The Bush Administration has not stated an official position on these types of proposals [17].

Wetland Restoration and Mitigation

Federal wetland policies during the past decade have increasingly emphasized restoration of wetland areas. Much of this restoration occurs as part of efforts to mitigate the loss of wetlands at other sites. The mitigation concept has broad appeal, but implementation has left a conflicting record. Examination of this record, presented in a June 2001 report from the National Research Council, found it to be wanting. The NRC report said that mitigation projects called for in permits affecting wetlands were not meeting the federal government's "no net loss" policy goal for wetlands function [18]. Likewise, a 2001 GAO report criticized the ability of the Corps to track the impact of projects under its current mitigation program that allows in-lieu-fee mitigation projects in exchange for issuing permits allowing wetlands development [19]. Both scientists and policymakers debate whether it is possible to restore or create wetlands with ecological and other functions equivalent to or better than those of natural wetlands that have been lost over time. Results so far seem to vary, depending on the type of wetland and the level of commitment to monitoring and maintenance. Congress has repeatedly endorsed mitigation in recent years.

Much of the attention to wetland restoration has focused on Louisiana, where an estimated 80% of the total loss of U.S. coastal wetlands has occurred (coastal wetlands are about 5% of all U.S. wetlands). The current rate of loss is more than 15,000 acres per year, a decline from higher rates in earlier years [20]. In response to these losses, Congress authorized a task force, led by the Corps, to prepare a list of coastal wetland restoration projects in the state, and also provided funding to plan and carry out restoration projects in this and other coastal states under the Coastal Wetlands Planning, Protection and Restoration

Act of 1990, also known as the Breaux Act. By 2006, 138 projects have been approved. Of this total, the completed projects have reestablished more than 32,000 acres, protected more than 38,000 acres, and enhanced (specific wetland functions have been intensified or improved) more than 320,000 acres. The remaining projects, when constructed, will establish or protect an additional 33,000 acres and enhance almost 195,000 acres. The completed projects have cost about $625 million and the remaining projects have a total estimated cost of more than $913 million [21].

In the wake of hurricanes Katrina and Rita, multiple legislative proposals have been introduced to fund additional restoration projects that have already been planned by the U.S. Army Corps of Engineers and to explore other opportunities that would restore and stabilize additional wetlands. More specifically, before the hurricanes, Congress was considering legislation that would have provided about $2 billion to the restoration effort. Since the hurricanes, more expansive options costing up to $14 billion that were proposed in the 1998 report, *Coast 2050*, are also being considered [22].

Many federal agencies have been active in wetland improvement efforts in recent years. In particular, the Fish and Wildlife Service (FWS) has been promoting the success of its Partners for Wildlife program. According to the program website, visited on July 14, 2005, the program had entered into almost 29,000 agreements with landowners to protect or restore about 640,000 acres of wetlands and more than 4,700 miles of riparian and in-stream habitat (and more than 1 million acres of upland habitat also) through FY2002. The website appears to include only data on FY2006 accomplishments [23].

Other programs also restore and protect domestic and international wetlands. One of these derives from the North American Wetlands Conservation Act, reauthorized through FY2007 in P.L. 107-304 with an appropriations ceiling that is increasing from $55 million in FY2003 to $75 million in FY2007. The act provides grants for wetland conservation projects in Canada, Mexico, and the United States. According to the FWS FY2007 budget notes, the United States. and its partners have protected more than 18.5 million acres and restored, created, or enhanced an additional 5.9 million acres through almost 1,500 projects. The FWS has combined funding for this program with several other laws into what it calls the North American Wetlands Conservation Fund.

Under the Convention on Wetlands of International Importance, more commonly known as the Ramsar Convention, the United States is one of 134 nations that have agreed to slow the rate of wetlands loss by designating important sites. These nations have designated 1,229 sites since the convention was adopted in 1971. The United States has designated 19 wetlands, encompassing 3 million acres.

Mitigation also has become an important cornerstone of the Section 404 program in recent years. A 1990 MOA signed by the agencies with regulatory responsibilities outlines a sequence of three steps leading to mitigation: first, activities in wetlands should be avoided when possible; second, when they can not be avoided, impacts should be minimized; and third, where minimum impacts are still unacceptable, mitigation is appropriate. It directs that mitigated wetland acreage be replaced on a one-for-one functional basis. Therefore, mitigation may be required as a condition of a Section 404 permit.

Some wetland protection advocates are critical of mitigation, which they view as justifying destruction of wetlands. They believe that the Section 404 permit program should be an inducement to avoid damaging wetland areas. These critics also contend that adverse impacts on wetland values are often not fully mitigated and that mitigation measures, even if well-designed, are not adequately monitored or maintained. Supporters of current efforts counter that they generally work as envisioned, but little data exist to support this view. Questions about implementation of the 1990 MOA and controversies over the feasibility of compensating for wetland losses further complicate the wetland protection debate.

In response to criticism in the NRC and GAO reports (discussed above), in November 2001, the Corps issued new guidance to strengthen the standards on compensating for wetlands lost to development. The guidance was criticized by environmental groups and some Members of Congress for weakening rather than strengthening mitigation requirements and for the Corps' failure to consult with other federal agencies. In December 2002, the Corps and EPA released an action plan including 17 items that both agencies believe will improve the effectiveness of wetlands restoration efforts [24].

In March, 2006, the Corps and EPA released a draft mitigation rule to replace the 1990 MOA with clearer requirements on what will be considered a successful project to compensate for wetlands lost to development or agriculture. The agencies identify the three purposes of these revisions as: improving the effectiveness of mitigation in replacing lost wetland functions and areas; expanding public participation in decision-making; and increasing the efficiency and predictability of both the mitigation process and the approval of mitigation banks. The rule was developed in response to a provision in the 2003 defense authorization bill (P.L. 107-314) that directed the Corps to establish mitigation project performance standards by 2005. Environmental activists fear that the rule will be even less protective than current policy. The comment period has been extended, and ends on June 30 [25].

The concept of "mitigation banks," in which wetlands are created, restored, or enhanced in advance to serve as "credits" that may be used or acquired by permit applicants when they are required to mitigate impacts of their activities, is widely endorsed. Numerous public and private banks have been established, but many believe that it is too early to assess their success. In its recent study of mitigation, the Environmental Law Institute determined that as of 2005, there were 330 active banks, 75 sold out banks, and 169 banks seeking approval to operate [26]. Provisions in several laws, such as the 1996 farm bill and the 1998 Transportation Equity Act (TEA-21), endorse the mitigation banking concept [27]. In November 2003, Congress enacted wetlands mitigation provisions as part of the FY2004 Department of Defense authorization act (P.L. 108-136).

FOR ADDITIONAL READING

Kusler, Jon and Teresa Opheim. *Our National Wetland Heritage: A Protection Guide.* Environmental Law Institute. [Washington] 1996. 149 pp.

National Academy of Sciences, National Research Council. *Compensating for Wetland Losses Under the Clean Water Act.* [Washington] 2001. 267 pp.

Strand, Margaret N. *Wetlands Deskbook.* Environmental Law Institute. [Washington] 1993. 883 pp.

U.S. Department of Agriculture, Economic Research Service. *Wetlands and Agriculture: Private Interests and Public Benefits,* by Ralph Heimlich et al. [Washington] 2001, 123 pp. Agricultural Economic Report No. 765.

U.S. Department of the Interior. U.S. Fish and Wildlife Service. *Status and Trends of Wetlands in the Coterminous United States 1998-2004.* [Washington] 2006. 54 pp.

U.S. General Accountability Office. *Wetlands Protection: Assessments Needed to Determine the Effectiveness of In-Lieu-Fee Mitigation.* (GAO-01-325) [Washington] May 2001. 75 pp.

—— *Waters and Wetlands: Corps of Engineers Needs to Evaluate District Office Practices in Determining Jurisdiction.* (GAO-04-297) [Washington] February 2004. 45 pp.

REFERENCES

[1] For additional information, see CRS Report RS22276, *Coastal Louisiana Ecosystem Restoration After Hurricanes Katrina and Rita,* by Jeffrey Zinn.

[2] See [http://www.whitehouse.gov/news/releases/2004/04/20040422-1.html].

[3] Office of the President, Council on Environmental Quality, *Conserving America'sWetlands 2006: Two Years of Progress Implementing the President's Goal,* April 2006, 47p.

[4] U.S. Fish and Wildlife Service, National Wetlands Inventory, *Status and Trends of Wetlands in the Coterminus United States, 1998 - 2004,* March 2006, 110 pp. This is the most recent of several status and trend reports by the Inventory over the past 25 years, which document wetlands trends at both a national and regional scale.

[5] U.S. Army, Corps of Engineers, "Regulatory Statistics, All Permit Decisions, FY2003." See [http://www.usace.army.mil/inet/functions/cw/cecwo/reg/2003webcharts.pdf].

[6] For additional information, see CRS Report RL31411, *Controversies over Redefining "Fill Material" Under the Clean Water Act,* by Claudia Copeland.

[7] For more information, see CRS Report 97-223, *Nationwide Permits for Wetlands Projects: Issues and Regulatory Developments,* by Claudia Copeland.

[8] For background, see CRS Report RS21421, *Mountaintop Mining: Background on Current Controversies,* by Claudia Copeland.

[9] For additional information, see CRS Report RL30849, *The Supreme Court Addresses Corps of Engineers Jurisdiction Over 'Isolated Waters': The SWANCC Decision,* by Robert Meltz and Claudia Copeland.

[10] See [http://www.epa.gov/owow/wetlands/guidance/SWANCC/index.html].

[11] U.S. Government Accountability Office, *Corps of Engineers Needs to Evaluate Its District Office Practices in Determining Jurisdiction,* GAO-04-297, February 2004, 45 pp.

[12] U.S. Congress, House of Representatives, Committee on Transportation and Infrastructure, Subcommittee on Water Resources and Environment, *Inconsistent Regulation of Wetlands and Other Waters,* Hearing 108-58, 108th Cong., 2d sess., Mar. 30, 2004.

[13] For additional information, see CRS Report RL33263, *The Wetlands Coverage of the Clean Water Act is Revisited by the Supreme Court: Rapanos and Carabell,* by Robert Meltz and Claudia Copeland.

[14] Natural Resources Conservation Service, *National Resources Inventory; 2002 Annual NRI (Wetlands).* At [http://www.nrcs.usda.gov/technical/land/nri02/nrio2

wetlands.hmtl].

[15] For more information on these provisions, see CRS Report RL31486, *Resource Conservation Title of the 2002 Farm Bill: A Comparison of New Law with Bills Passed by the House and Senate, and Prior Law*; and for the status of implementation, see the 2002 farm bill implementation subsection of CRS Issue Brief IB96030, *Soil and Water Conservation Issues*, both by Jeffrey A. Zinn.

[16] U.S. Congress, House of Representatives, Committee on Transportation and Infrastructure, Subcommittee on Water Resources and Environment, *The Wetland Permitting Process: Is It Working Fairly?* Hearing 107-50, 107th Cong., 1st sess., Oct. 3, 2001.

[17] For more information, see CRS Report RL30423, *Wetlands Regulation and the Law of Property Rights "Takings"*, by Robert Meltz.

[18] National Academy of Sciences, National Research Council, *Compensating for Wetland Losses under the Clean Water Act* (Washington, DC: 2001), 267 pp.

[19] U.S. Government Accountability Office, *Wetlands Protection: Assessments Needed to Determine the Effectiveness of In-Lieu-Fee Mitigation*, GAO-01-325, 75 pp.

[20] Loss rates have been calculated by U.S. Geological Survey's Nation Wetlands Research Center, which has published a number of reports describing past and predicted loss rates.

[21] Louisiana Coastal Wetlands Conservation and Restoration Task Force, *Coastal Wetlands Planning, Protection, and Restoration Act (CWPPRA): A Response to Louisiana's Wetland Loss*, 2006, 16 pp.

[22] For a more detailed discussion of the effects of the hurricanes on planning for wetland restoration, see CRS Report RS22276, *Coastal Louisiana Ecosystem Restoration After Hurricanes Katrina and Rita*, by Jeffrey Zinn.

[23] See [http://www.ecos.fws.gov/partners], visited June 12, 2006.

[24] U.S. Environmental Protection Agency and U.S. Army Corps of Engineers, "National Wetlands Mitigation Action Plan, Dec. 24, 2002." See [http://www.epa.gov/owow/wetlands/ pdf/map1226withsign.pdf].

[25] Information on compensatory mitigation can be found at [http://www.epa.gov/ wetlandsmitigation].

[26] For more information on mitigation generally, and mitigation banks specifically, see Environmental Law Institute, *2005 Status Report on Compensatory Mitigation in the United States*, April 2006, 105 pp.

[27] For more information on the early history of banking, see CRS Report 97-849, *Wetland Mitigation Banking: Status and Prospects*, by Jeffrey A. Zinn.

In: Environmental Research Advances
Editor: Peter A. Clarkson, pp. 95-116

ISBN: 978-1-60021-762-3
© 2007 Nova Science Publishers, Inc.

Chapter 4

THE USE OF CONSTRUCTED WETLANDS FOR WASTEWATER TREATMENT IN THE CZECH REPUBLIC

Jan Vymazal

ENKI o.p.s, Trebon, Czech Republic and Duke University Wetland Center,
Nicholas School of the Environment and Earth Sciences,
Durham, NC, USA

ABSTRACT

The first experiments on the use of wetland macrophytes to treat wastewaters were carried out in the Czech Republic as early as in 1970s but the first full-scale constructed wetland (CW) for wastewater treatment was built only in 1989. The last survey recognized 155 systems in operation at the end of 2002. With few exceptions all systems are designed with sub-surface horizontal flow. Most systems (91) were designed for the treatment of municipal or domestic sewage while 56 CWs were designed for the treatment of wastewater from combined sewer systems. Size distribution of constructed wetlands varies between 3 and 1200 PE (population equivalent) with most systems being designed for the size between 100 and 500 PE (60 systems) followed by small on-site systems for less than 10 PE (43 systems). As most CWs are designed with the specific area of about 5 m^2 per one population equivalent the most frequent size of vegetated beds is between 500-2500 m^2 (67 systems) and < 50 m^2 (42 systems). The area of vegetated beds varies between 9 and 5630 m^2. The most frequently used macrophyte is Common reed (*Phragmites australis*) which is used in 42 systems as a monotypic stand and in 50 systems in combination with other macrophytes such as Reed canarygrass *(Phalaris arundinacea)* or Cattails (*Typha* spp.). Constructed wetlands are very effective in removing organics BOD_5, COD, suspended solids and bacteria. Removal of BOD_5 and suspended solids is usually > 90% and removal of bacteria is commonly > 99%. Removal of nutrients is lower not exceeding 50% for both nitrogen and phosphorus. Constructed wetlands in the Czech Republic proved to be a suitable alternative to conventional treatment systems for small sources of pollution.

1. INTRODUCTION

Wetlands have been intensively studied in the Czech Republic for more than three decades. The early studies were carried out during the International Biological Program (IBP) for Wetland Management at the Institute of Botany of Academy of Sciences of the Czech Republic (formerly Czechoslovakia) at Třeboň in southern Bohemia and in Brno in southern Moravia. The experiments carried out during the IBP were aimed primarily at wetland ecology, ecophysiology of wetland plants (primary productivity, biomass, evapotranspiration, mineral nutrition and role of plants in nutrient cycling) and the role of algae in wetlands. The results were summarized in several publications (Dykyjová and Květ 1978).

In late 1960's and early 1970's, numerous estimates of biomass, production, and nutrient content in helophyte communities proved their manifold connections with the availability of nutrients. The experiments were carried out in various natural wetlands including heavily polluted sites near wastewater treatment plant outfall (e.g. Dykyjová 1971a, 1971b, Dykyjová and Květ 1982, Dykyjová and Hradecká 1973, Dykyjová et al. 1970, Fiala and Květ 1971, Květ 1971, 1973, 1975, Květ and Ondok 1971, Ondok 1970). The most measurements were done in stands of Common reed (*Phragmites australis* (Cav.) Trin. ex Steud., formerly *Phragmites communis* Trin.), however, other macrophytes were also investigated, e.g. *Schoenoplectus lacustris* (= *Scirpus lacustris*), *Glyceria maxima* (= *G. aquatica*), *Sparganium erectum*, *Bolboschoenus maritimus*, *Acorus calamus*, *Typha latifolia*, *T. angustifolia* and *Phalaris arundinacea*.

In order to verify data from natural stands experimental cascade hydroponic system was constructed at Třeboň. The experiments have been carried out for three years in a defined nutrient medium (Dykyjová 1978, Dykyjová et al. 1972). Unfortunately, the results of these experiments were not used for any kind of wastewater treatment design. The interest in the use of wetland macrophytes for wastewater treatment was revived in the Czech Republic only at the end of 1980s. After a period of experiments in a small-scale constructed wetland (Vymazal 1990) the first full-scale constructed wetland for the treatment of runoff waters from a dung-hill was built in 1989. Due to lack of rainwater and thus runoff in the summer of 1989 it has been decided to use the system for the treatment of sewage from the adjacent village. Despite the fact the treatment system was built with little knowledge of constructed wetlands and was originally designed for different type of wastewater, the treatment effect was very high (Vymazal 1998). However, the appearance of the system, which was really far from neat, became a pretence for negative opinions on constructed wetlands given by various organizations (hygiene service, water inspection, the Ministry of the Environment etc.). Unfortunately, the results themselves were not taken into consideration and, as a result, in 1990 none and in 1991 only four constructed wetlands for wastewater treatment were built and put in full operation.

Since 1992 a steep increase in a number of constructed wetlands has appeared (Fig. 1). The major factor influencing this phenomenon was the cancellation of "recommended list of treatment systems for small point sources of pollution" at the end of 1991. This list offered various technologies (e.g. oxidation ditch, rotating biological contactors (RBCs) or simple activated sludge technologies) but did not include constructed wetlands (however, it did include soil filtration). Another reason for the fact that constructed wetlands were built in the Czech Republic after 1991 was the change in the socio-economic sphere. The towns and villages became much more independent - they handled their own budget and the decision-making rights became larger (Vymazal 1998).

2. RESULTS OF 2003 SURVEY

2.1. Number and Types of Systems

The survey carried out in 2003 revealed that by the end of 2002 at least 155 constructed wetlands for wastewater treatment were in operation (Fig. 1). The number is quite exact for larger systems, i.e. systems > 50 PE (population equivalent), but the number of small on-site systems may be larger because these small systems are difficult to track. The largest number built in a single year was 23 in 1995 (Figure 1).

All constructed wetlands in the Czech Republic are designed with sub-surface horizontal flow (Figure 2). It is called horizontal flow because the wastewater is fed in at the inlet and flows slowly through the porous medium under the surface of the bed in a more or less horizontal path until it reaches the outlet zone where it is collected before leaving via level control arrangement at the outlet. During this passage the wastewater comes into contact with a network of aerobic, anoxic and anaerobic zones.

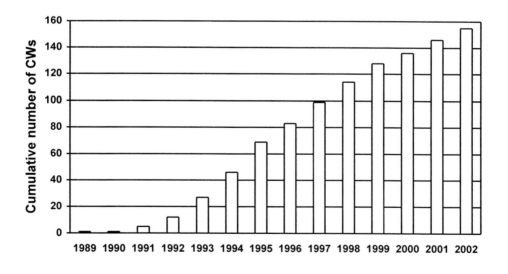

Figure 1. Number of constructed wetlands for wastewater treatment in the Czech Republic.

2.2. Type of Wastewater and Pretreatment

Most constructed wetlands in the Czech Republic are designed to treat municipal or domestic sewage (91 systems) or sewage combined with stormwater runoff (56 systems). Other types of wastewater include those from dairies, bakeries, abattoir facilities and goat farm. One CW is a part of the system designed for the treatment of landfill leachate and one

CW has been designed to treat stormwater runoff only. With the exception of 6 CWs all systems have been designed as a secondary treatment step.

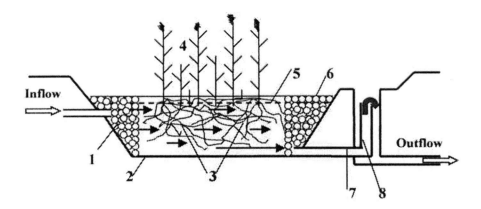

Figure 2. Schematic layout of the sub-surface horizontal flow constructed wetland. 1-distribution zone filled with large stones, 2-impermeable liner (usually plastic liner), 3-filtration medium (gravel, crushed rock), 4-vegetation, 5-water level in the bed, 6-collection zone, 7-collection pipe, 8-water level adjustment structure.

CWs with sub-surface flow require high level of pretreatment which should remove primarily suspended solids as clogging of the bed is the major threat for a proper functioning of the system. Small systems (ca. < 50 PE) usually use a septic tank, larger systems use Imhoff tanks. If the system is designed to treat wastewater from combined sewer system (i.e,. in combination with stormwater runoff) also a grit chambers or sand traps are included.

2.3. Size of Constructed Wetlands

The size of constructed wetlands according to population equivalent (PE) ranges between 3 and 1200. However, most systems (60) were designed to treat wastewater from sources of pollution between 101-500 PE. Recently, number of on-site CWs for single households has increased and at present at least 43 systems for less than 10 PE are in operation.

Surface area of all secondary systems in the Czech Republic has been designed so far according to a simple formula (Kickuth 1977, Cooper, 1990; Vymazal, 1998; Vymazal et al., 1998b):

$$A_h = Q_d(\ln C_o - \ln C) / K_{BOD},$$

where A_h = surface area of bed (m^2)
$\quad Q_d$ = average flow (m^3 d^{-1})
$\quad C_o$ = influent BOD$_5$ (mg L^{-1})
$\quad C$ = effluent BOD$_5$ (mg L^{-1})
$\quad K_{BOD}$ = rate constant (m d^{-1})

Formerly proposed value of K_{BOD} 0.19 m d^{-1} by Kickuth (1977) resulted in too small area of the bed and consequently lower treatment effect. At present, the value of K_{BOD} is usually set at 0.1 m d^{-1} according to the recommendation of Cooper (1990). Despite some criticism this value proved to be sufficient in case the system is designed to remove organics (BOD$_5$ and COD) and suspended solids (Vymazal 2001). For mechanically pretreated domestic and municipal sewage this generally means that the specific area is about 5 m^2 PE^{-1}. The survey revealed that the average specific area of the Czech constructed wetlands designed for secondary treatment of sewage (n = 138) is 5.1 m^2 PE^{-1} (range: 2.0 – 13.3 m^2 PE^{-1}) with 74 systems having the specific area between 4.0 and 6.0 m^2 PE^{-1}. Specific area of systems designed for tertiary treatment (n = 6) ranges between 0.27 and 2.0 m^2 PE^{-1}. However, it should be noted that instead of PE, number of connected people better fits the reality. In the design, number of connected people equals PE which is set at 60 g BOD$_5$ d^{-1} person^{-1} but at present, one person usually produces only 30 - 40 g BOD$_5$ per day.

The area of vegetated beds varies between 9 and 5 630 m^2 with 67 systems having bed area between 501 and 2 500 m^2 and 42 systems having bed area < 50 m^2. Small and on-site systems usually use only one bed. Systems with the bed surface area larger than about 300 m^2 use more beds, mostly 2 to 4. These systems are often built with the possibility of both parallel or series configuration.

Aspect ratio (i.e., length:width) varies between 0.33 and 5.3 with majority of systems built with aspect ratio < 2 and many of them have aspect ratio < 1 (Vymazal 1998). The major reason for low aspect ratio is the need of wastewater distribution to as much wide profile as possible in order to avoid clogging of the inlet zone because clogging of the inlet zone and subsequent surface flow may deteriorate treatment effect of the process.

2.4. Filtration Media

Filtration media in constructed wetlands with horizontal sub-surface flow should 1) facilitate macrophyte growth, 2) provide high and sustainable filtration effect, and 3) maintain high hydraulic conductivity. In the beginning of the use constructed wetlands with horizontal sub-surface flow heavy soils were commonly used. The filtration and consequently removal effect of this medium was very high but clogging occurred very often after a short period of operation. In late 1980s, more coarse filtration materials were proposed (Cooper 1990). Czech constructed wetlands use, with few exceptions, coarse media – gravel, crushed rock or gravel-sand. The early systems usually used fractions 4-8 mm. At present, gravel or crushed rock fraction 8-16 mm is often preferred because it has been shown that this fraction provides sufficient hydraulic conductivity while supporting a healthy macrophyte growth and good treatment efficiency.

2.5. Vegetation

The most important effects of macrophytes in relation to treatment process in sub-surface horizontal flow CWs are the physical effects of the plant tissue (i.e., erosion control, filtration effect, provision of surface area for attached microorganisms, insulation of the bed surface). The metabolism of the macrophytes (plant uptake, oxygen release and release of antibiotics) is of the lesser importance in the HSF CWs (Vymazal et al. 1998a). More than one decade of the experience has shown that under the climatic conditions of the Czech Republic the most important role of plants in CWs with sub-surface horizontal flow is the insulation of the bed during the winter period (Figure 3).

Figure 3. Reed bed at Slavošovice in January 2004. Standing litter together with a snow layer provide a good insulation during the winter.

The most frequently used plant in the Czech HSF CWs is *Phragmites australis* (Common reed) (Fig. 4) which is used either singly (42 systems) or in combination (50 systems) with other macrophyte species such as *Phalaris arundinacea* (Reed canarygrass, Figure 4) and *Typha latifolia* (Common cattail). Other species used singly are *Phalaris arundinacea*, *Glyceria maxima* (Mannagrass), and *Typha latifolia*. The early systems mostly used rhizomes from natural localities. However, this method did not provide rapid growth of vegetation. It also introduced many weed species and in addition, the planting was limited due to climatic conditions to the period of May-August. At present, most systems are planted with seedlings grown in a nursery from seeds. The use of seedlings enables the planting of beds from early April to late October for *Phalaris* and from late April to late September for *Phragmites* (Vymazal, 1998). This method is very successful and provides fast coverage of the bed surface. This is especially true for *Phalaris* which creates a dense cover even during the first growing season and reaches its maximum biomass during the second growing season. The density 4-8 seedlings per 1 m^2 is sufficient. Weeds, if present, are restricted to vegetated bed margins. Vegetation is not harvested in most cases. The major reason for that is that litter provides an excellent insulation of the surface of vegetated beds during periods of cold weather.

Figure 4. Constructed wetland at Slavošovice (left) planted with *Phragmites australis* and constructed wetland at Břehov (right) planted with combination of *Phalaris arundinacea* (sides) and *P. australis* (between stripes of *Phalaris*).

Figure 5. Mixture of *Iris pseudacorus* and *I. sibirica* in the system at Žitenice.

Small on-site systems very often use combination of decorative plants (Fig. 5), such as *Iris pseudacorus* (Yellow flag), *Iris sibirica* (Blue flag), *Filipendula ulmaria* (Queen of the meadow), *Epilobium hirsutum* (Hairy willow-herb) or *Lythrum salicaria* (Purple loosestrife).

2.6. Treatment Efficiency

Constructed wetlands with horizontal sub-surface flow have been shown very effective in removing organics and suspended solids (Kadlec and Knight 1996, Vymazal et al. 1998b). Removal of nutrients is lower and usually does not exceed 50% in municipal or domestic wastewater.

2.6.1. Removal of Organics

Organic compounds are degraded aerobically as well as anaerobically by bacteria attached to plant underground organs (i.e., roots and rhizomes) and filtration media surface. The oxygen required for aerobic degradation is supplied directly from the atmosphere by diffusion or oxygen leakage from the macrophyte roots and rhizomes in the rhizosphere. Aerobic degradation of soluble organic matter is governed by the aerobic heterotrophic bacteria according to the following reaction:

$$(CH_2O) + O_2 \rightarrow CO_2 + H_2O \tag{1}$$

Cooper et al. (1996) pointed out that also ammonifying bacteria degrade organic compounds containing nitrogen under aerobic conditions. Both bacterial groups consume organics but the faster metabolic rate of the heterotrophs means that they are mainly responsible for the reduction in the BOD_5 of the system. Insufficient supply of oxygen to this group will greatly reduce the performance of aerobic biological oxidation, however, if the oxygen supply is not limited, aerobic degradation will be governed by the amount of active organic matter available to the organisms. In most systems designed for the treatment of domestic or municipal sewage the supply of dissolved organic matter is sufficient and aerobic degradation is limited by dissolved oxygen concentration. On the other hand, besides hetrotrophic and ammonifying bacteria also nitrifying bacteria utilize oxygen to cover their physiological needs. However, it is generally agreed that heterotrophic bacteria outcompete nitrifying bacteria for oxygen (e.g., Brix 1998).

Anaerobic respiration occurs in the soil zone below the Fe^{3+} reduction zone. The process can be carried out in wetland soils by either facultative or obligate anaerobes. It represents one of the major ways in which high molecular weight carbohydrates are broken down to low molecular weight organic compounds, usually as dissolved organic carbon, which are, in turn, available to microbes. Anaerobic degradation is a multi-step process. In the first step the primary end-products of fermentation are fatty acids such as acetic (Equation 2), butyric, and lactic (Equation 3) acids, alcohols (Equation 4) and the gases CO_2 and H_2 (Mitsch and Gosselink 1986, Reddy and Graetz 1988, Vymazal 1995, Vymazal et al. 1998a):

$$C_6H_{12}O_6 \rightarrow 3\ CH_3COOH + H_2 \tag{2}$$

$$C_6H_{12}O_6 \rightarrow 2\ CH_3CHOHCOOH\ (lactic\ acid) \tag{3}$$

$$C_6H_{12}O_6 \rightarrow 2\ CO_2 + 2\ CH_3CH_2OH\ (ethanol) \tag{4}$$

Acetic acid is the primary acid formed in most flooded soils and sediments. Strictly anaerobic sulfate-reducing bacteria (Equation 5) and methane-forming bacteria (Equations 6 and 7) then utilize the end-products of fermentation and, in fact, depend on the complex community of fermentative bacteria to supply substrate for their metabolic activities. Both groups play an important role in organic matter decomposition and carbon cycling in wetland soil environments (Grant and Long 1985, Vymazal 1995):

$$CH_3COOH + H_2SO_4 \rightarrow 2\ CO_2 + 2\ H_2O + H_2S \tag{5}$$

$$CH_3COOH + 4\,H_2 \rightarrow 2\,CH_4 + 2\,H_2O \qquad\qquad (6)$$

$$4\,H_2 + CO_2 \rightarrow 2\,CH_4 + 2\,H_2O \qquad\qquad (7)$$

Methane is an important gas evolving from flooded and saturated soils. The major pathways of methane formation are 1) decarboxylation of acetic acid at the expense of hydrogen (Equation 6) and 2) reduction of electron acceptors (Equation 7) or by a unique aceticlastic reaction (Grant and Long 1985):

$$CH_3COOH \rightarrow CH_4 + CO_2 \qquad\qquad (8)$$

The acid-forming bacteria are fairly adaptable but the methane-formers are more sensitive and will only operate in the pH range 6.5 to 7.5. Over-production of acid by the acid-formers can rapidly result in a low pH value, thus stopping the action of the methane-forming bacteria and resulting in production of odorous compounds from the constructed wetland. Anaerobic degradation of organic compounds is much slower than aerobic degradation. However, when oxygen is limiting at high organic loadings, anaerobic degradation will predominate (Cooper et al. 1996).

Most constructed wetlands for wastewater treatment in the Czech Republic are designed with the primary aim to remove organics (BOD_5 and COD). The results of the survey indicated that removal or organic compounds is very high (Table 1).

Table 1. Removal of organics in Czech constructed wetlands. Concentrations in mg L^{-1}, loadings in kg ha^{-1} d^{-1}. IN=inflow, OUT=outflow, EFF=removal efficiency in % (based on individual efficiencies), n=number of annual means, N=number of systems

	Concentrations					Loadings				
Parameter	IN	OUT	EFF	n	N	IN	OUT	REM	n	N
BOD_5	153	14.8	85.6	175	63	41.4	6.0	35.4	79	29
COD	334	54	75.6	106	38	99	21.3	77.7	51	25

Removal of both BOD_5 and COD is high and outflow concentrations are very low. However, an average BOD_5 concentration is comparable with mean outflow concentrations of 10.3 mg L^{-1} and 18.3 mg L^{-1} reported for CWs with horizontal sub-surface flow from Denmark (Schierup et al. 1990) and North America (Kadlec and Knight 1996), respectively. There is considerably less information on COD values but Schierup et al. (1990) reported a mean COD outflow concentration in Danish systems 64.7 mg L^{-1}. Data presented in Figure 6 show one of the advantages of constructed wetlands as compared to conventional treatment systems. Constructed wetlands do not require certain minimum inflow concentration of organics (BOD_5 ca. 50-80 mg L^{-1}) and treat successfully diluted (e.g. with stormwater) wastewater. Also, constructed wetlands successfully handle fluctuations in inflow concentrations and the treatment performance does not depend on the season.

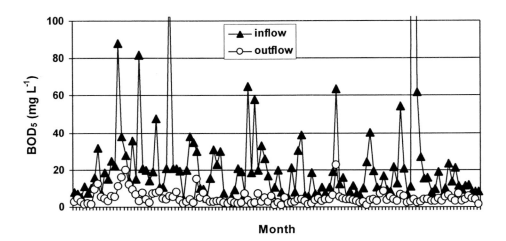

Figure 6. Removal of BOD₅ in CW at Spálené Poříčí during the period November 1992 – December 2002. Samples taken on a monthly basis.

Table 2. Average inflow (IN), outflow (OUT) and removed (REM) mass loading rates (kg ha^{-1} d^{-1}) of BOD₅ and COD in vegetated beds of CWs with horizontal sub-surface flow. N=number of systems

BOD₅	IN	OUT	REM	N
North America[1]	33.2	12.5	20.7	16
Poland[2]	49.3	9.5	30.8	4
Denmark[3]	44.8	7.9	36.9	54
Slovenia[4]	63.4	16.7	46.7	6
Sweden[5]	22.4	3.1	19.3	2
COD	IN	OUT	REM	N
Poland[2]	114	35.4	78.6	4
Denmark[3]	127	30.2	96.9	48
Slovenia[4]	244	114	130	10

[1]Kadlec and Knight (1996), [2]Kowalik and Obarska-Pempkowiak (1998)
[3]Schierup et al. (1990), [4]Urbanc-Berčič et al. (1998), [5]Sundblad (1998)

The average inflow BOD₅ loading of the Czech constructed wetlands was 41.4 (±35.4) kg ha^{-1} d^{-1} with variation between 2.6 and 203 kg ha^{-1} d^{-1}. The average outflow BOD₅ loading was 6.0 (± 7.6) kg ha^{-1} d^{-1} and varied between 0.14 and 77 kg ha^{-1} d^{-1}. The average removed BOD₅ loading was 35.4 (± 37.7) kg ha^{-1} d^{-1}. All the numbers are within the range reported in the literature (Table 2). The average inflow COD loading was 99 (± 89) kg ha^{-1} d^{-1} with variation between 11.8 and 474 kg ha^{-1} d^{-1}. The average outflow COD loading was 21.3 (± 24.8) kg ha^{-1} d^{-1} and varied between 3.1 and 96 kg ha^{-1} d^{-1}. The average removed COD loading was 77.7 (± 77.4) kg ha^{-1} d^{-1}. The average COD loadings of the Czech constructed wetlands are somewhat lower as compared to literature data but still within the reported range. The levels of both BOD₅ and COD removed loadings are well predictable (Figure 7).

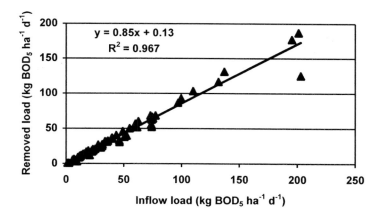

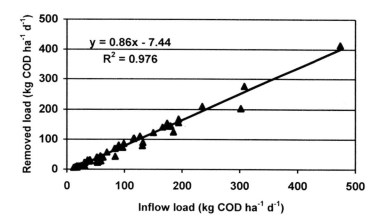

Figure 7. Relationship between inflow and outflow BOD_5 and COD loadings in the Czech constructed wetlands.

2.6.2. Removal of Suspended Solids

Suspended solids are effectively removed in constructed wetlands by filtration and sedimentation. For CWs with horizontal sub-surface flow it is very important to include reliable pretreatment units as excessive concentrations of suspended solids may clog filtration beds. Effective units when properly operated and maintained may remove up to 60% of the inflow concentration.

Analysis of 114 annual means from 41 systems obtained from the survey revealed that average inflow and outflow concentrations of total suspended solids (TSS) were 165 mg L^{-1} and 12.6 mg L^{-1}, respectively. The mean treatment effect based on individual treatment efficiencies was 84.7%. In general, various types of constructed wetlands provide very low outflow TSS concentrations but CWs with horizontal sub-surface flow are probably the most effective. Kadlec and Knight (1996) reported mean outflow TSS concentration for North American CWs with horizontal sub-surface flow 10.3 mg L^{-1} and Brix (1994) reported an average TSS outflow concentration of 13.6 mg L^{-1} for 77 systems in Denmark and U.K.

An average inflow TSS load of 40.5 ($\pm$ 45.8) kg ha^{-1} d^{-1} is reduced to 10.0 ($\pm$ 13.7) kg ha^{-1} d^{-1}. Similar results were reported by Brix (1994) for 51 Danish systems – average inflow load of 52.2 ($\pm$ 63.7) kg ha^{-1} d^{-1} was reduced to 10.6 ($\pm$ 15.0) kg ha^{-1} d^{-1}. Removal of TSS mass load is highly correlated with the inflow TSS load ($R^2 = 0.995$).

2.6.3. Removal of Nitrogen

Nitrogen is removed from wastewater in CWs with horizontal sub-surface flow via nitrification/denitrification, adsorption and plant uptake with nitrification/denitrification being the major processes responsible from nitrogen retention/removal. Volatilisation of ammonia does not take place in these systems because there is no free water surface. Due to prevailing anoxic or/and anaerobic conditions in the filtration bed, the removal of nitrogen is limited by low nitrification potential.

Nitrification is usually defined as the biological oxidation of ammonium to nitrate with nitrite as an intermediate in the reaction sequence. This definition has some limitations where heterotrophic microorganisms are involved but is adequate for the autotrophic and dominant species (Hauck 1984). Nitrification has been typically associated with the chemoautotrophic bacteria, although it is now recognized that heterotrophic nitrification occurs and can be of significance (Keeney 1973, Paul and Clark 1996). Nitrification can occur in the surface aerobic layer and in the root zone. The diffusion of oxygen from wetland plant roots to adjacent soil creates an aerobic environment around the roots. This aerobic soil envelope around the roots can support nitrification of NH_4^+ diffused from surrounding anaerobic zones.

Nitrification rate in wetland soils depends on the supply of NH_4^+ to the aerobic zone, pH and alkalinity of water, temperature, moisture, inorganic C source, presence of nitrifying bacteria and the thickness of the aerobic soil layer (Vymazal 1995). Nitrification is a chemoautotrophic process. The nitrifying bacteria derive energy from the oxidation of ammonia and/or nitrite and carbon dioxide is used as a carbon source for synthesis of new cells. These organisms require O_2 during ammonium-N oxidation to nitrite-N and nitrite-N oxidation to nitrate N (Eqs.9, 10 and 11). Oxidation of ammonium to nitrate is a two-step process (Hauck 1984):

$$NH_4^+ + 1.5\,O_2 \rightarrow NO_2^- + 2\,H^+ + H_2O \tag{9}$$

$$NO_2^- + 0.5\,O_2 \rightarrow NO_3^- \tag{10}$$

$$NH_4^+ + 2\,O_2 \rightarrow NO_3^- + 2\,H^+ + H_2O \tag{11}$$

The first step, the oxidation of ammonium to nitrite, is executed by strictly chemolithotrophic (strictly aerobic) bacteria which are entirely dependent on the oxidation of ammonia for the generation of energy for growth. In soil, species belonging to the genera *Nitrosospira* (*N. briensis*). *Nitrosovibrio* (*N. tenuis*), *Nitrosolobus* (*N. multiformis*), *Nitrosococcus* (*N. nitrosus*) and *Nitrosomonas* (*N. europaea*) have been identified. *N. europaea* is also found in fresh waters (Grant and Long 1985). The chemoautotrophic nitrifiers are generally aerobes that derive their C largely from CO_2 or carbonates. The demonstration of an anaerobic metabolism in *Nitrosomonas europaea* using pyruvate as an electron donor and NO_2^- as an electron acceptor in the presence of NH_4^+, however, shows the physiological diversity of these organisms (Paul and Clark 1996). The probable reaction sequence for the oxidation of ammonia to nitrite by Nitroso group bacteria is (Hauck 1984):

Ammonia $\Rightarrow$ *hydroxylamine* $\Rightarrow$ *nitroxyl* $\Rightarrow$ *nitrohydroxylamine* $\Rightarrow$ *nitrite*

The second step in the process of nitrification, the oxidation of nitrite to nitrate, is performed by facultative chemolitrotrophic bacteria which can also use organic compounds, in addition to nitrite, for the generation of energy for growth. In contrast with the ammonia-oxidizing bacteria, only one species of nitrite-oxidizing bacteria is found in the soil and freshwater, *i.e.*, *Nitrobacter winogradskyi* (Grant and Long 1985). Paul and Clark (1996) reported, in addition to *Nitrobacter*, also a genus *Nitrospira* (*N. gracilus, N. marina*) to be found in marine environments while Schmidt (1982) reported that genus *Nitrospira* may be also found in soil and freshwater environments. Also, in contrast to ammonia-oxidizing bacteria, nitrite-oxidizing bacteria, or at least some species, can grow mixotrophically on nitrite and a carbon source, or are even able to grow in the absence of oxygen (Bock *et al.* 1986).

Denitrification is most commonly defined as the biological reduction of NO_2^- or NO_3^- to N_2 or gaseous N oxides (Hauck 1984, Paul and Clark 1996). However, the assimilatory reduction of NO_3^- to NH_4^+ and nitrification also produce N oxides (N_2O and/or NO), so that a more precise definition is desirable to keep pace with current knowledge (Hauck 1984). From a biochemical viewpoint, denitrification is a bacterial process in which nitrogen oxides (in ionic and gaseous forms) serve as terminal electron acceptors for respiratory electron transport. Electrons are carried from an electron-donating substrate (usually, but not exclusively, organic compounds) through several carrier systems to a more oxidized N form. The resultant free energy is conserved in ATP, following phosphorylation, and is used by the denitrifying organisms to support respiration. Denitrification is illustrated by following equation (Hauck 1984):

$$6\,(CH_2O) + 4\,NO_3^- \Rightarrow 6\,CO_2 + 2\,N_2 + 6\,H_2O \qquad (12)$$

This reaction is irreversible, and occurs in the presence of available organic substrate only under anaerobic or anoxic conditions (Eh = +350 to +100 mV), where nitrogen is used as an electron acceptor in place of oxygen. More and more evidence is being provided from pure culture studies that nitrate reduction can occur in the presence of oxygen. Hence, in waterlogged soils nitrate reduction may also start before the oxygen is depleted (Kuenen and Robertson 1987, Laanbroek 1990). Gaseous N production during denitrification can also be depicted by (Hauck 1984):

$$4\,(CH_2O) + 4\,NO_3^- \rightarrow 4\,HCO_3^- + 2\,N_2O + 2\,H_2O \qquad (13)$$

$$5\,(CH_2O) + 4\,NO_3^- \rightarrow H_2CO_3 + 4\,HCO_3^- + 2\,N_2 + 2\,H_2O \qquad (14)$$

Diverse organisms are capable of denitrification. In an array are organotrophs, lithotrophs, phototrophs, and diazotrophs (Paul and Clark 1996). Most denitrifying bacteria are chemoheterotrophs. They obtain energy solely through chemical reactions and use organic compounds as electron donors and as a source of cellular carbon (Hauck 1984). The genera *Bacillus*, *Micrococcus* and *Pseudomonas* are probably the most important in soils; *Pseudomonas*, *Aeromonas* and *Vibrio* in the aquatic environment (Grant and Long 1981). A list of genera involved in the denitrification process has been given by *e.g.*, Focht and Verstraete (1977) or Paul and Clark (1996). When oxygen is available, these organisms oxidize a carbohydrate substrate to CO_2 and H_2O (Reddy and Patrick 1984). Aerobic respiration using oxygen as an electron acceptor or anaerobic respiration using nitrogen for this purpose is accomplished by the denitrifiers with the same series of electron transport system. This facility to function both as an aerobe and as an anaerobe is of great practical importance because it enables denitrification to proceed at a significant rate soon after the onset of anoxic conditions (a redox potential of about 300 mV) without change in microbial population (Hauck 1984). Because denitrification is carried out almost exclusively by facultative anaerobic heterotrophs that substitute oxidized N forms for O_2 as electron acceptors in respiratory processes, and because these processes follow aerobic biochemical routes, it can be misleading to refer to denitrification as an anaerobic process. It is rather one that takes place under anoxic conditions (Hauck 1984). It is generally agreed that the actual sequence of biochemical changes from nitrate to elemental gaseous nitrogen is:

$$2\,NO_3^- \Rightarrow 2\,NO_2^- \Rightarrow 2\,NO^- \Rightarrow N_2O^- \Rightarrow N_2 \qquad (15)$$

Vymazal (1995) summarized that environmental factors known to influence denitrification rates including absence of O_2, redox potential, soil moisture, temperature, pH value, presence of denitrifiers, soil type, organic matter, nitrate concentration and the presence of overlying water.

Nitrification and denitrification are known to occur simultaneously in flooded soils and sediments where both aerobic and anaerobic zones exist (Patrick and Reddy 1976). Flooded soil or sediments containing an aerobic surface layer over an anaerobic layer or the aerobic root rhizosphere of a wetland plant growing in an anaerobic soil could be the examples. By combining these two reactions, a balanced equation occurring in aerobic and anaerobic layers can be written as (Patrick and Reddy 1976):

$$24\,NH_4^+ + 48\,O_2 \Rightarrow 24\,NO_3^- + 24\,H_2O + 48\,H^+ \qquad (16)$$

$$24\,NO_3^- + 5\,C_6H_{12}O_6 + 24\,H^+ \Rightarrow 12\,N_2 + 30\,CO_2 + 42\,H_2O \qquad (17)$$

$$24\,NH_4^+ + 5\,C_6H_{12}O_6 + 48\,O_2 \Rightarrow 12\,N_2 + 30\,CO_2 + 66\,H_2O + 24\,H^+ \qquad (18)$$

Ionized ammonia may be adsorbed from solution through a cation exchange reaction with detritus, inorganic sediments or soils. The adsorbed ammonia is bound loosely to the substrate and can be released easily when water chemistry conditions change. At a given ammonia concentration in the water column, a fixed amount of ammonia is adsorbed to and saturates the available attachment sites. When the ammonia concentration in the water column is reduced (*e.g.*, as a result of nitrification), some ammonia will be desorbed to regain the equilibrium with the new concentration. If the ammonia concentration in the water column is increased, the adsorbed ammonia also will increase. If the wetland substrate is exposed to oxygen, perhaps by periodic draining, sorbed ammonium may be oxidized to nitrate (Kadlec and Knight 1996). Ammonium ion (NH_4^+) is generally adsorbed as an exchangeable ion on clays, and chemosorbed by humic substances, or fixed within the clay lattice. However, constructed wetlands with horizontal sub-surface flow use gravel or crushed rock and these media provide only limited number of sorption sites.

Nitrogen assimilation refers to a variety of biological processes that convert inorganic nitrogen forms into organic compounds that serve as building blocks for cells and tissues. The two forms of nitrogen generally used for assimilation are ammonia and nitrate nitrogen. Because ammonia nitrogen is more reduced energetically than nitrate, it is preferable source of nitrogen for assimilation. Although ammonia is the preferred nitrogen source, nitrate can also be used by some plant species (Kadlec and Knight 1996).

Biota utilize nitrate and ammonium, and decomposition processes release organic nitrogen and ammonium back to the water. In temperate climates, macrophyte uptake is a spring-summer phenomenon. Both above- and belowground plant parts grow during this period, but death phenomena are different. Plants such as *Typha* spp. or *Phragmites australis* in northern climates have an obvious annual cycle of aboveground biomass: new shoots start from zero biomass in early spring and grow at a maximum rate in spring and early summer. Late summer is a period of reduced growth, and complete shoot die back occurs in the fall. Vegetation nutrient concentrations tend to be highest early in the growing season, decreasing as the plant mature and senesce. Patterns of seasonal changes in composition vary for both the species and nutrients and broad generalization probably cannot be made (Vymazal 1995). As the rate of biomass and nutrient accumulation diminishes, translocation of nutrients and photoassimilate from leaves to rhizome occurs. The rates vary among species and are influenced by many environmental factors and may reach even more than 50% (Vymazal 1995).

The potential rate of nutrient uptake by plant is limited by its net productivity (growth rate) and the concentration of nutrients in the plant tissue. Nutrient storage is similarly dependent on plant tissue nutrient concentrations, and also on the ultimate potential for biomass accumulation: that is, the maximum standing crop. Therefore, desirable traits of a plant used for nutrient assimilation and storage would include rapid growth, high tissue nutrient content, and the capability to attain a high standing crop (biomass per unit area) (Reddy and DeBusk 1987). However, the field experience has shown that the amount of nitrogen removed from sewage via harvesting of macrophytes is usually negligible in temperate climate regions (Vymazal 2001).

The survey revealed (Table 3) that the mean inflow total nitrogen of 56 mg L^{-1} was reduced to 27.6 mg L^{-1} with an average removal efficiency of 47%. The average outflow concentration is within the range reported in the literature for treatment of municipal sewage in CWs with horizontal sub-surface flow. Schierup et al. (1990), Kowalik and Obarska-Pempkowiak (1998) and Sundblad (1998) reported an average outflow TN concentration from Danish, Polish and Swedish systems 20.9 mg L^{-1}, 24.5 mg L^{-1} and 15.1 mg L^{-1}, respectively. The removal of ammonia-N is lower (Table 3) and amounted to only 33.3% with an average outflow concentration of 18.4 mg L^{-1}. Schierup et al. (1990) reported an average outflow ammonia-N concentration of 14.1 mg L^{-1} and removal effect of 32.9% for 52 Danish systems. CWs with horizontal sub-surface flow do not provide high level of nitrification because of lack of oxygen in the filtration bed and, therefore, low nitrification rate is the major limiting factor in total nitrogen removal. This is especially important for sewage treatment because ammonia is the major form of nitrogen in sewage. In addition, sub-surface systems provide high rate of ammonification, i.e., reduction of organic compounds containing nitrogen to ammonia. This process proceeds under both aerobic and anaerobic conditions. As a result, ammonification increases ammonia concentration in the wastewater during the treatment process. The survey revealed that the average inflow concentration of organic-N 16.3 mg L^{-1} was reduced to only 2.7 mg L^{-1} in the outflow. On the other hand, horizontal sub-surface flow systems provide suitable conditions for denitrification – the average outflow nitrate-N concentration was only 2.45mg L^{-1}. Denitrification potential of CWs with horizontal sub-surface flow is limited by low nitrification during which ammonia-N is oxidized to nitrate-N and low concentration of nitrate in sewage.

Table 3. Removal of nitrogen in Czech constructed wetlands. Concentrations in mg L^{-1}, loadings in g m^{-2} yr^{-1}. IN=inflow, OUT=outflow, EFF=removal efficiency in % (based on individual efficiencies), n=number of annual means, N=number of systems

Parameter	Concentrations					Loadings				
	IN	OUT	EFF	n	N	IN	OUT	REM	n	N
Tot-N	56	27.6	47.0	37	16	819	467	352	49	19
NH_4^+-N	28.1	18.4	33.3	73	30	513	318	199	54	23

Removal of inflow nitrogen load was 43% based on the data from 19 systems. The average inflow to vegetated beds 819 g m^{-2} yr^{-1} is higher than loadings reported in the literature (Table 4). However, the removal effect is comparable with literature data.

Table 4. Average inflow, outflow and removed mass loading rates (g m^{-2} yr^{-1}) for total nitrogen in vegetated beds of CWs with horizontal sub-surface flow. EFF=removal effect (%), N=number of systems

	Average loading				
	Inflow	Outflow	Removed	EFF	N
[1]Austria	385	121	264	68.6	3
[2]Denmark	344	204	140	40.7	51
[3]North America	482	268	214	44.3	12
[4]Poland	577	456	121	20.9	6
[5]Sweden	577	318	259	44.9	3

[1]Haberl and Perfler (1990), [2]Schierup et al. (1990), [3]Kadlec and Knight (1996).
[4]Kowalik and Obarska-Pempkowiak (1998), [5]Sundblad (1998).

2.6.4. Removal of Phosphorus

Phosphorus is retained in wetlands in a long-term basis by soil accretion and adsorption, precipitation, plant uptake (with plant harvest) and detritus sorption. The specific conditions of CWs with horizontal sub-surface flow (no free water surface and consequently no contact of wastewater with litter or accreted layer) limit removal mechanisms to adsorption, precipitation and plant uptake.

Phosphorus is adsorbed on soil or sediment particles if sufficient Al, Fe, Ca and Mg are present. Which ion is more active in sequestering P depends on the pH of the system and the amount of the ion present (Richardson 1999). In acidic soils, inorganic P is adsorbed on hydrous oxides of Fe and Al, and P may precipitate as insoluble Fe-phosphates (Fe-P) and Al-P. Precipitation as insoluble Ca-P and Mg-P is the dominant transformation at pH's greater than 8.0. Adsorption of phosphorus is greater in mineral vs. organic soils (Richardson 1999).

The chemical processes of soil adsorption and precipitation are considered more important than uptake by plants and detritus, although rates would vary considerably among wetlands (Richardson and Craft 1993). A general comparison of the amount of P adsorption among wetlands soil types demonstrates that wetland systems vary greatly in terms of the amount of P that can be removed by this process. The systems with the highest adsorption capacity have been shown to have mineral soils, which contain the highest amount of Fe or Al (Richardson 1985). Assessment of desorption rates are also required to determine net storage via this mechanism (Richardson 1999). One of the proposed mechanisms for the release of phosphorus from soils upon submergence is the reductive dissolution of Fe(III) and Mn(IV) phosphate minerals (Patrick et al. 1973).

Phosphorus is removed primarily by ligand exchange reactions, where phosphate displaces water or hydroxyls from the surface of Fe and Al hydrous oxides. However, media used for CWs with horizontal sub-surface flow, i.e. pea gravel or crushed rock, usually do not contain great quantities of Fe, Al or Ca and therefore, removal of phosphorus is generally low in these systems. Also, removal of phosphorus via plant harvesting is negligible for wetlands designed to treat sewage (Vymazal 2001).

Removal of phosphorus in Czech constructed wetlands is low and amounted to only 41.3% based on the results from 25 systems. The mean inflow concentration of 6.85 mg L^{-1} was reduced to 3.3 mg L^{-1} at the outflow. The outflow concentration is within the range reported for constructed wetlands designed for sewage treatment. Kadlec and Knight (1996), Kowalik and Obarska-Pempkowika (1998), Börner et al. (1998) and Brix (1994) reported an average outflow concentrations of 2.97 mg L^{-1}, 4.10 mg L^{-1}, 3.99 mg L^{-1} and 6.3 mg L^{-1} for North American, Polish, German and Danish/UK systems, respectively. Both inflow and outflow loading rates recorded in the survey for the Czech constructed wetlands were within the range reported in the literature (Table 5).

Table 5. Average inflow, outflow and removed mass loading rates (g m^{-2} yr^{-1}) for total phosphorus in vegetated beds of CWs with horizontal sub-surface flow. EFF=removal effect (%), N=number of systems

| | Average loading | | | | |
	Inflow	Outflow	Removed	EFF	N
[1]Denmark + UK	120	95	25	20.8	50
[2]New Zealand	262	201	61	23.3	2
[3]N. America	188	146	42	22.3	8
[4]Poland	99	58	41	41.4	5
[5]Sweden	148	57	91	61.5	3
This study	118	75	43	36.4	21

3. CASE STUDIES

3.1. On-Site System Žitenice

Constructed wetland at Žitenice (Figure 8) was built in 1994 for a single house. It is dimensioned for 4 PE with the vegetated bed area of 18 m^2. Filtration material is coarse sand. In the beginning of operation, vegetation cover was *Phragmites australis* and *Typha latifolia* but recently, the owner replaced these two species with decorative plants with *Iris pseudacorus* being the dominant species. For pretreatment, an upgraded septic tank is used. An average flow through the system was 0.7 m^3 d^{-1}.

Figure 8. Constructed wetland at Zitenice.

Table 6. Treatment performance of the constructed wetland at Žitenice during the period January 2002-December 2003

Parameter	Inflow (mg L^{-1})	After pretreatment (mg L^{-1})	Outflow (mg L^{-1})	Efficiency (%)
BOD$_5$	348	74.2	8.4	97.6
COD	1074	270	33	96.9
TSS	593	41.3	6.2	99.0
TP	16.3	8.5	6.4	60.7
TN	73.7	65	38.8	47.4

In Table 6, treatment performance of the system during the period January 2002 – December 2003 is presented. The data indicate an excellent removal of organics (BOD$_5$ and COD) and total suspended solids (TSS) with all parameters being removed with more than 95%. Removal of phosphorus is higher than usual level achieved in CWs with horizontal sub-surface flow and amounted to 60.7% probably due to higher sorption capacity of sand which is used as filtration medium. In general, sand provides more sorption sited than gravel or crushed rock. Removal of nitrogen amounted to the level typical for this type of constructed wetlands achieved in the Czech Republic. The inflow total-N concentration is quite high but typical for undiluted domestic sewage. When evaluating the treatment efficiency of this system, it is important to underline a very high treatment efficiency of pretreatment unit.

3.2. Constructed Wetland at Čistá

Constructed wetland at Čistá (Figure 9) which is one of the largest systems in the Czech Republic was built in 1995 and is designed to treat wastewater from a combined sewer system. Pretreatment consists of screen bars and Imhoff tank. Total vegetated beds area of 3 040 m^2 is divided in two parallel parts both consisting of two beds in series. All four beds equal in size 760 m^2. The system was designed for 800 PE and in 2003 789 people were connected to the treatment plant. Crushed rock (fraction 8-12 mm) was used for filtration medium. Vegetation cover consists of stripes of *Phragmites australis* and *Phalaris arundinacea* planted perpendicular to wastewater flow. The average flow of wastewater is 180 m^3 d^{-1}.

Figure 9. Constructed wetland at Čistá.

The results (Table 7) indicate very high and consistent removal effect for all monitored parameters. The discharge limits are set at 15 mg L^{-1} for both BOD_5 and suspended solids and 50 mg L^{-1} for COD.

Table 7. Treatment performance of constructed wetland at Čistá. Mean annual concentrations in mg L^{-1}, removal effect (Eff) in %

	BOD$_5$			COD			TSS		
	In	Out	Eff	In	Out	Eff	In	Out	Eff
1995	34	5.7	83.2	108	23	78.7	26.3	1.3	95.1
1996	35	7.3	79.1	100	43	57.0	96	4.5	95.3
1997	43	9.2	78.6	113	39	65.5	33	4.0	87.9
1998	75	4.9	93.5	207	37	82.1	39	3.5	91.0
1999	24	5.0	79.2	95	33	65.3	15	1.7	89.7
2000	45	5.9	86.9	133	38	71.4	85	5.2	93.9
2001	34	3.9	88.5	125	28	77.6	85	5.0	94.1
2002	53	5.1	90.4	140	34	75.7	15	6.3	58.0
2003	33	6.2	81.2	105	23	78.1	24	5.0	79.2

4. CONCLUSIONS

Constructed wetlands with horizontal sub-surface flow have been used for nearly 15 years in the Czech Republic. By the end of 2002 at least 155 systems were put in operation mostly for secondary treatment of domestic or municipal sewage. Removal of organics (BOD$_5$ and COD) and suspended solids is very high and steady over the years of operation. Removal of nutrients (nitrogen and phosphorus) is usually low and does not exceed 50% for municipal sewage when systems are dimensioned at about 5 m^2 per population equivalent. Nitrogen removal is limited by lack of oxygen in filtration bed and consequently low nitrification while phosphorus removal is limited by low sorption capacity of filtration materials (gravel, crushed rock). However, constructed wetlands with horizontal sub-surface flow proved to be a viable alternative for wastewater treatment for small sources of pollution especially when organics and suspended solids are the treatment target.

REFERENCES

Börner, T.; von Felde, K.; Gschlössl, T.; Kunst, S.; Wissing, F.W. In *Constructed Wetlands for Wastewater Treatment in Europe*; Vymazal, J.; Brix, H.; Cooper, P.F.; Green, M.B.; Haberl, R.; Eds.; Backhuys Publishers: Leiden, The Netherlands; 1998; pp 169-190.

Brix, H. In *Global Wetlands. Old World and New*; Mitsch, W.J.; Ed.; Elsevier: Amstedam, The Netherlands, 1994; pp 325-333.

Brix, H. In *Constructed Wetlands for Wastewater Treatment in Europe*; Vymazal, J.; Brix, H.; Cooper, P.F.; Green, M.B.; Haberl, R.; Eds.; Backhuys Publishers: Leiden, The Netherlands, 1998; pp 123-152..

Cooper, P.F. *European Design and Operations Guidelines for Reed Bed Treatment Systems*; Water Research Centre: Swindon, U.K., 1990; pp 1-33.

Cooper, P.F.; Job, G. D.; Green, M. B.; Shutes, R. B. E. *Reed Beds and Constructed Wetlands for Wastewater Treatment*; WRc Publications: Medmenham, U.K.,1996; pp 1-184.

Dykyjová, D. *Photosynthetica* 1971a, *5, 329-340.*

Dykyjová, D. *Hidrobiol.* 1971b, *12, 361-376.*

Dykyjová, D. In *Pond Littoral Ecosystems. Structure and Functioning*; Dykyjová, D.; Květ, J.; Eds.; Springer Verlag: Berlin, Germany, 1978; pp 257-277.

Dykyjová, D.;Hradecká, D. *Pol. Arch. Hydrobiol.* 1973, *20, 111-119.*

Dykyjová, D.; Květ, J. (Eds.) *Pond Littoral Ecosystems. Structure and Functioning.* Springer Verlag, Berlin, Germany, 1978; pp.1-283.

Dykyjová, D.; Květ, J. 1982. In *Wetland Ecology and Management;* Gopal, B.; Turner, R.E.; Wetzel, R.G.; Whigham, D.F.; Eds.; Scientific Publications: Jaipur, India, 1982, pp. 335-355.

Dykyjová, D.; Ondok, J.P.; Pribán, K. *Photosynthetica* 1970, *4, 280-287.*

Dykyjová, D.; Ondok, J.P.; Hradecká, D. *Folia Geobot. Phytotax.* 1972, *7, 259-268.*

Fiala, K.; Květ, J. In *The Scientific Management of Animal and Plant Communities for Conservation;* Duffey, E.; Watt, A.S.; Eds.; Blackwell Scientific Publications: Oxford, UK, 1971; pp 241-269.

Grant, W. D.; Long, P.E. In *The Handbook of Environmental Chemistry. Vol. 1, Part D. The Natural Environmental and Biogeochemical Cycles*; Hutzinger, O.; Ed.; Springer Verlag: Berlin, Germany, 1985; pp 125-237.

Haberl, R.; Perfler, R. In *Constructed Wetlands in Water Pollution Control*; Cooper, P.F.; Findlater, B.C., Ed.; Pergamon Press: Oxford, U.K., 1990; pp 205-214.

Hauck, R.D. In *The Handbook of Environmental Chemistry. Vol. 1., Part C, The Natural Environment and Biogeochemical Cycles*; Hutzinger, O.; Ed.; Springer-Verlag: Berlin, Germany, 1984; pp 105-127.

Kadlec, R.H.; Knight, R.L. *Treatment Wetlands*; Lewis Publishers: Boca Raton, FL, 1996; 1-893.

Keeney, D.R. *J. Environ. Qual.* 1973, *2, 15-29.*

Kickuth, R. In *Utilization of Manure by Land Spreading*; Comm. Europ. Commun, EUR 5672e, London, U.K., pp 335-343.

Kowalik, P.; Obarska-Pempkowiak, H. In *Constructed Wetlands for Wastewater Treatment in Europe*; Vymazal, J.; Brix, H.; Cooper, P.F.; Green, M.B.; Haberl, R.; Eds.; Backhuys Publishers: Leiden, The Netherlands, 1998; 217-225.

Kuenen, J.G.; Robertson, L.A. In *The Nitrogen and Sulphur Cycles*; Cole, J.A.; Ferguson, S.J.; Eds.; Cambridge University Press: Cambridge, UK, 1987; pp 162-218.

Laanbroek, H.J. *Aquat. Bot.* 1990, *38, 109-125.*

Květ, J. *Hidrobiol.* 1971, *12, 14-40.*

Květ, J. *Arch. Hydrobiol.* 1973, *20, 137-147.*

Květ, J. *Symp. Biol. Hungar.* 1975, *15, 113-123.*

Květ, J.; Ondok, J.P. *Photosynthetica* 1971, *5, 417-420.*

Mitsch, W.J.; Gosselink, J.G. *Wetlands*; Van Nostrand Reinhold Company: New York, USA, 1986; pp 1-670.

Ondok, J.P. *Preslia* 1970, *42, 256-261.*

Patrick, W.H., Jr.; Reddy, K.R. *J. Environ. Qual.* 1976, *5, 469-472.*

Patrick, W.H., Jr.; Gotoh, S.; Williams, B.G. *Science* 1973, *179, 564-566.*

Paul, E.A.; Clark, F.E. *Soil Microbiology and Biochemistry*; 2nd ed; Academic Press: San Diego, CA, 1996; pp 339.

Reddy, K. R., DeBusk, W. F. In *Aquatic Plants for Water Treatment and   Resource Recovery*; Reddy, K.R.; Smith, W.H.; Eds.; Magnolia Publishing: Orlando, FL, 1987; pp 337-357.

Reddy, K.R.; Graetz, D.A. In *Ecology and Management of Wetlands*. Vol. 1. *Ecology of Wetlands*; Hook, D.D. et al.; Eds.; Timber Press: Portland, OR, 1988; pp 307-318.

Richardson, C.J. *Science* 1985, *228, 1424-1427*.

Richardson, C.J. In . In *Phosphorus Biogeochemistry in Subtropical Ecosystems*; Reddy, K.R.; O'Connor, G.A.; Schelske, C.L.; Eds.; CRC Press: Boca Raton, Florida, 1999. pp 47-68.

Richardson, C.J.; Craft, B.C. In *Constructed Wetlands for Water Quality Improvement*; Moshiri, G.A.; Eds.; Lewis Publishers: Boca Raton, FL, 1993; pp 271-282.

Schierup, H.-H.; Brix, H.; Lorenzen, B. *Spildevandsrensning i rodzoneanlæg. Status for danske anlæg 1990 samt undersøgelse og vurdering af de vigtigste renseprocesser. (Wastewater treatment in reed beds);* Aarhus University, Denmark,1990; pp 1-87.

Schmidt, E.L. In *Nitrogen in Agricultural Soil*; Stevenson, F.J.; Ed.; Am. Soc. Agron.: Madison, WI, 1982; pp 253-267.

Sundblad, K. In *Constructed Wetlands for Wastewater Treatment in Europe*; Vymazal, J.; Brix, H.; Cooper, P.F.; Green, M.B.; Haberl, R.; Eds.; Backhuys Publishers: Leiden, The Netherlands, 1998; pp 251-259.

Urbanc-Berčič, O.; Bulc, T.; Vrhovšek, D. In *Constructed Wetlands for Wastewater Treatment in Europe*; Vymazal, J.; Brix, H.; Cooper, P.F.; Green, M.B.; Haberl, R.; Eds.; Backhuys Publishers: Leiden, The Netherlands, 1998; 241-250.

Vymazal, J. In *Constructed Wetlands in Water Pollution Control*; Cooper, P.F.; Findlater, C.B.; Eds.; Pergamon Press: Oxford, UK, 1990; pp. 347-358.

Vymazal, J. *Algae and Element Cycling in Wetlands*; Lewis Publishers: Chelsea, MI, 1995; pp 1-689.

Vymazal, J. In *Constructed Wetlands for Wastewater Treatment in Europe*; Vymazal, J.; Brix, H.; Cooper, P.F.; Green, M.B.; Haberl, R.; Eds.; Backhuys Publishers: Leiden, The Netherlands, 1998; pp 95-121.

Vymazal, J. In *Transformations of Nutrients in Natural and Constructed Wetlands;* Vymazal, J.; Ed.; Backhuys Publishers: Leiden, The Netherlands, 2001; pp 1-93.

Vymazal, J.; Brix, H.; Cooper, P.F.; Haberl, R.; Perfler, R.; Laber, J. In *Constructed Wetlands for Wastewater Treatment in Europe*; Vymazal, J.; Brix, H.; Cooper, P.F.; Green, M.B.; Haberl, R.; Eds.; Backhuys Publishers: Leiden, The Netherlands, 1998a; pp 7-66.

Vymazal, J.; Brix, H.; Cooper, P.F.; Green, M.B.; Haberl, R. Eds.; *Constructed Wetlands for Wastewater Treatment in Europe*; Backhuys Publishers: Leiden The Netherlands, 1998b; pp 1-366.

In: Environmental Research Advances
Editor: Peter A. Clarkson, pp. 117-134

ISBN: 978-1-60021-762-3
© 2007 Nova Science Publishers, Inc.

Chapter 5

NATIONAL PARK MANAGEMENT[*]

Carol Hardy Vincent, Susan Boren and Sandra L. Johnson

ABSTRACT

The 109th Congress is considering legislation and conducting oversight on National Park Service (NPS) related topics. The Administration is addressing park issues through budgetary, regulatory, and other actions. Earlier Congresses and Administrations also have dealt with similar issues. While this report focuses on several key topics, others may be added if circumstances warrant.

Historic Preservation. The NPS administers the Historic Preservation Fund (HPF), which provides grants to states and other entities to protect cultural resources. Congress provides annual appropriations for the HPF, and views differ as to whether to retain the federal role in financing the fund. Legislation to reauthorize the HPF (S. 1378 and H.R. 5861) is being considered. Further, the Advisory Council on Historic Preservation has issued a draft revision of its policy statement regarding treatment of burial sites, and the draft has been controversial.

Maintenance Backlog. Attention has focused on the NPS's maintenance backlog, estimated by DOI at between $5.80 billion and $12.42 billion for FY2005. Views differ as to whether the backlog has increased or decreased in recent years, and the NPS has been defining and quantifying its maintenance needs. H.R. 1124 and S. 886 seek to eliminate the NPS maintenance backlog and the annual operating deficit.

Policy Revisions. The NPS has revised its service-wide management policies — one of the authorities governing decision-making on a wide range of issues. The final policies, issued August 31, 2006, dropped many of the proposed changes that were controversial. The House and Senate have held hearings on this issue, related NPS authorities, and broader management issues.

Wild and Scenic Rivers. The Wild and Scenic Rivers System preserves free-flowing rivers, which are designated by Congress or through state nomination with Secretarial approval. The NPS, and other federal agencies with responsibility for managing designated rivers, prepare management plans to protect river values. Management of lands within river corridors is sometimes controversial, because of a variety of issues

---

[*] Excerpted from CRS Report RL33483, June 20, 2006.

including the possible effects of designation on private lands and of corridor activities on the rivers. Legislation is pending to designate, study, or extend components of the system, and some of these measures have passed the Senate or House.

Other Issues. Some other park management topics of interest to the 109th Congress are covered here. They relate to the competitive sourcing initiative, whereby certain NPS activities judged to be commercial in nature are subject to public-private competition; air quality at national park units; and security of park units, particularly at national icons and along international borders.

INTRODUCTION

The National Park System is perhaps the federal land category best known to the public. The National Park Service (NPS) in the Department of the Interior (DOI) manages 390 units, including 58 units formally entitled *national parks* and a host of other designations [1]. The system has more than 84 million acres [2]. The NPS has an appropriation of about $2.28 billion for FY2006. As of January 10, 2006, the agency employed 24,679 federal employees and used an additional 137,000 volunteers. An estimated 263 million people visited park units in 2004.

The NPS statutory mission is multifaceted: to conserve, preserve, protect, and interpret the natural, cultural, and historic resources of the nation for the public, and to provide for their use and enjoyment by the public. The use and preservation of resources has appeared to some as contradictory and has resulted in management challenges. Attention centers on how to balance the recreational use of parklands with the preservation of park resources, and determine appropriate levels and sources of funding to maintain NPS facilities and to manage NPS programs. In general, activities that harvest or remove resources from units of the system are not allowed. The NPS also supports the preservation of natural and historic places and promotes recreation outside the system through grant and technical assistance programs.

The establishment of several national parks preceded the 1916 creation of the National Park Service (NPS) as the park system management agency. Congress established the nation's first national park — Yellowstone National Park — in 1872. The park was created in the then-territories of Montana and Wyoming "for the benefit and enjoyment of the people," and placed "under the exclusive control of the Secretary of the Interior" (16 U.S.C. §§21-22). In the 1890s and early 1900s, Congress created several other national parks mostly from western public domain lands, including Sequoia, Yosemite, Mount Rainier, Crater Lake, and Glacier. In addition to the desire to preserve nature, there was interest in promoting tourism. Western railroads, often recipients of vast public land grants, were advocates of many of the early parks and built grand hotels in them to support their business.

There also were efforts to protect the sites and structures of early Native American cultures and other special sites. The Antiquities Act of 1906 authorized the President to proclaim national monuments on federal lands that contain "historic landmarks, historic and prehistoric structures, and other objects of historic or scientific interest" (16 U.S.C. §431). Most national monuments are managed by the NPS. (For more information, see CRS Report RS20902, *National Monument Issues*, by Carol Hardy Vincent).

There was no system of national parks and monuments until 1916, when President Wilson signed a law creating the NPS to manage and protect the national parks and many of the monuments. That *Organic Act* provided that the NPS "shall promote and regulate the use of the Federal areas known as national parks, monuments, and reservations ... to conserve the scenery and the natural and historic objects and the wild life therein and to provide for the enjoyment of the same in such manner and by such means as will leave them unimpaired for the enjoyment of future generations" (16 U.S.C. §1). President Franklin D. Roosevelt greatly expanded the system of parks in 1933 by transferring 63 national monuments and historic military sites from the USDA Forest Service and the War Department to the NPS.

The 109th Congress is considering legislation or conducting oversight on many NPS-related topics. Several major topics are covered in this report: historic preservation through the Historic Preservation Fund, which is administered by the NPS; the NPS maintenance backlog; an NPS review of agency policies; and management of wild and scenic rivers, which are administered by the NPS or another land management agency. Other issues addressed in brief are activities of the NPS under the President's Competitive Sourcing Initiative, air quality at national park units, and security of NPS units and lands.

While in some cases the topics covered are relevant to other federal lands and agencies, this report does not comprehensively cover topics primarily affecting other lands/agencies. For background on federal land management generally, see CRS Report RL32393, *Federal Land Management Agencies: Background on Land and Resources Management*, coordinated by Carol Hardy Vincent. Overview information on numerous natural resource issues, focused on resource use and protection, is provided in CRS Report RL32699, *Natural Resources: Selected Issues for the 109th Congress*, coordinated by Carol Hardy Vincent, Nicole T. Carter, and Julie Jennings. Information on appropriations for the NPS is included in CRS Report RL33399, *Interior, Environment, and Related Agencies: FY2007 Appropriations*, coordinated by Carol Hardy Vincent and Susan Boren. Information on BLM and Forest Service lands is contained in CRS Report RL33596, *Federal Lands Managed by the Bureau of Land Management (BLM) and the Forest Service*, coordinated by Ross W. Gorte and Carol Hardy Vincent.

Several other NPS-related topics are not covered in this brief. Some of them, or other topics, may be added to this brief if events warrant. For example, how national park units are created and what qualities make an area eligible to be an NPS unit are of continuing interest. (For more information, see CRS Report RS20158, *National Park System: Establishing New Units*, by Carol Hardy Vincent.) Second, legislation has been considered in recent Congresses to study, designate, and fund particular National Heritage Areas (NHAs) as well as to establish a process and criteria for designating and managing NHAs. (For more information, see CRS Report RL33462, *Heritage Areas: Background, Proposals, and Current Issues*, by Carol Hardy Vincent and David Whiteman.) Third, recent decades have witnessed increased demand for a variety of recreational opportunities on federal lands and waters. New forms of motorized recreation have gained in popularity, and the use of motorized off-highway vehicles (OHVs) has been particularly contentious. (For more information, see CRS Report RL33525, *Recreation on Federal Lands*, coordinated by Kori Calvert and Carol Hardy Vincent.) Fourth, the management of the NPS concessions program, which provides commercial visitor services, continues to receive oversight. Finally, the role of gateway communities in NPS planning and the impact of land uses on gateway communities have received increased attention.

CURRENT ISSUES

Historic Preservation

Background. The National Historic Preservation Act of 1966 (NHPA; P.L. 89-665, 16 U.S.C. §479) created a program of state grants for historic preservation under the Historic Preservation Fund (HPF). The program has been expanded to include Indian tribal grants; grants for Alaska Natives and Native Hawaiians; restoration grants for buildings at historically black colleges and universities (HBCUs); and Save America's Treasures grants. The major purpose of the HPF program is to protect cultural resources.

Administered by the National Park Service, the HPF provides grants-in-aid to states and territories for activities specified in the NHPA. These grants are funded on a 60% federal/ 40% state matching share basis. States carry out program purposes directly through State Historic Preservation Offices or through subgrants and contracts with public and private agencies, organizations, institutions of higher education, and private individuals. Under current law, 10% of each state's annual allocation distributed by the Secretary of the Interior is to be transferred to local governments that are certified eligible under program regulation.

Some Members of Congress support proposals to eliminate a federal government role in financing the HPF, leaving such programs to be sustained by private support. A case in point is the National Trust for Historic Preservation, for which permanent federal funding was eliminated in FY1998. Others assert that a federal role in supporting historic preservation is necessary and should be maintained. One example of a program receiving bipartisan support is the Save America's Treasures program, currently funded under the HPF. The HPF, authorized by the National Historic Preservation Act Amendments of 2000 (NHPA; P.L. 106-208), expired at the end of FY2005 but has continued to be funded.

Administrative Actions. President Bush's annual budget requests, including the request for FY2007, have recommended funding for a Preserve America program (previously established by Executive Order 13287). The program consists of competitive grants providing one-time assistance to encourage community preservation of cultural, historic, and natural heritage through education and heritage tourism. It serves as an adjunct to Save America's Treasures. For FY2006, Congress provided that a portion of Save America's Treasures funds could be allocated to Preserve America's grants. The first round of Preserve America grants for FY2006 (totaling $3.5 million) was announced on March 9, 2006. Funds for Save America's Treasures were first appropriated in FY1999 and used to restore such historic documents as the Star Spangled Banner, the Declaration of Independence, and the U.S. Constitution. These projects require a 50% cost share, and no single project can receive more than one grant from this program.

The FY2007 Administration budget contained $71.9 million for the Historic Preservation Fund. It proposed shifting funding for National Heritage Areas to the HPF, as part of a new America's Heritage and Preservation Partnership Program. Funding for Heritage Partnerships was proposed to be cut from $13.3 million in FY2006 to $7.4 million for FY2007. The Save America's Treasures program would have been cut in half, from $29.6 million to $14.8 million, but the Preserve America grants would have doubled — from $5.0 million to $10.0 million. The establishment of the new program was not adopted by the House or the Senate

Committee on Appropriations in H.R. 5386, the Interior, Environment, and Related Agencies appropriations bill for FY2007. (See "Legislative Activity," below).

The Advisory Council on Historic Preservation (ACHP) was established as an independent agency by the NHPA to advise Congress and the President on historic preservation matters. The ACHP has issued a draft revision of its policy statement regarding treatment of burial sites (71 *Fed. Reg.* 13066, March 14, 2006). The policy guides federal agencies in making decisions about burial sites, human remains, and funerary objects encountered during reviews under §106 of the National Historic Preservation Act. Section 106 requires federal agencies to take into account the effects of their undertakings on historic properties. The ACHP is revising its policies on the grounds that the current one no longer reflects its position, because since its issuance in 1988 there have been changes in law and regulations affecting how human remains and funerary objects are to be considered and treated. The draft policy has been controversial. It brings to the forefront issues of the power of the ACHP, the extent to which Indian tribes in particular are being accommodated, and whether the ACHP can make or is making final decisions in §106 reviews. The draft policy was open for public comment through July 28, 2006, and the ACHP is currently assessing the comments received for the development of a final policy.

Legislative Activity. Most of the recent congressional action on historic preservation has been in appropriations, since the authorization typically has been for five-year periods (most recently through FY2005). P.L. 109-54 provided $72.2 million for HPF for FY2006. The FY2007 House-passed appropriation for HPF is $58.7 million, while the level reported by the Senate Appropriations Committee is $70.7 million. Both the House and the Senate Committee supported $35.7 million for grants-in-aid to states, $3.9 million for tribal grants, and $1.0 million for HBCUs. The House-passed bill contained $15.0 million for Save America's Treasures, and $3.0 million for Preserve America. The Senate Appropriations Committee-reported bill contained $30.0 million for Save America's Treasures, including not more than $10.0 million for Preserve America grants. (For more information on funding for historic preservation, see CRS Report RL33617, *Historic Preservation: Background and Funding*, by Susan Boren).

H.R. 5861 and S. 1378 seek to reauthorize the HPF (§108, NHPA) and to amend provisions pertaining to the operation of the Advisory Council on Historic Preservation. Both bills would extend authority to fund the HPF through FY2015 through annual deposits of $150 million earned from oil and gas development from the Outer Continental Shelf . They also would make changes to the membership and operation of the ACHP. Further, the House bill provides that a certified local government that uses an eligibility determination to initiate local regulatory requirements must allow due process to property owners who might object to an eligibility determination on their property. It also seeks to protect applicants who must submit to §106 review from being required to fund surveys and studies to determine potential effects on historic properties that are beyond their identified area. The Senate bill was reported from committee on April 20, 2006 (S.Rept. 109-235) and the House bill was passed by the House on September 25, 2006.

P.L. 109-234, the Emergency Supplemental Appropriations Act for Defense, the Global War on Terror, and Hurricane Recovery, includes $43.0 million for the HPF. Of those funds, $40.0 million would be to establish a specialized grants-in-aid program for the repair and rehabilitation of historic structures damaged by Hurricanes Katrina and Rita, particularly for those properties listed on the National Register of Historic Places. Grants would be directed to endangered historic properties in major disaster areas, including those within National Heritage Areas. A nonfederal match is not required, and no more than 5% of the funds may be used for administrative expenses. The remaining $3.0 million would be for §106 assistance. The President's FY2006 supplemental request and the House-passed bill would have provided $3.0 million for the HPF for rehabilitation of historic structures. The Senate-passed bill would have provided $83.0 million for the HPF, an amount that exceeded the FY2006 appropriation for all of HPF's programs ($72.2 million).

On September 20, 2006, a subcommittee of the House Committee on Government Reform held oversight hearings on the effects of historic preservation on economic and community development, and on the implementation of the federal historic rehabilitation tax credit program administered by the National Park Service in partnership with the Internal Revenue Service and state historic preservation officers. A focus was on the benefits of historic preservation in community development, and in particular on how the state and federal tax credit programs have worked in partnership to enhance rehabilitation of properties and communities. Since the inception of the federal rehabilitation tax credit program in 1976, over $36 billion in private investment in historic buildings has been generated, involving over 32,000 approved projects [3]. Some changes to this tax credit program are included in the Community Restoration and Revitalization Act (H.R. 3159/H.R. 659), which would allow an increased rehabilitation tax credit for certain low-income buildings. These bills have been referred to a House committee; there has been no further legislative action.

Maintenance Backlog

Background. The NPS has maintenance responsibility for buildings, trails, recreation sites, and other infrastructure. There is debate over the levels of funds to maintain this infrastructure, whether to use funds from other programs, and how to balance the maintenance of the existing infrastructure with the acquisition of new assets. Congress continues to focus on the agency's *deferred maintenance*, often called the *maintenance backlog* — essentially maintenance that was not done when scheduled or planned. DOI estimates deferred maintenance for the NPS for FY2005, based on varying assumptions, at between $5.80 billion and $12.42 billion with a mid-range figure of $9.11 billion. While the other federal land management agencies — the Forest Service (FS), Bureau of Land Management (BLM), and Fish and Wildlife Service (FWS) — also have maintenance backlogs, congressional and administrative attention has centered on the NPS backlog. For FY2005, the FS estimates its backlog at $5.97 billion, while DOI estimates the FWS backlog at between $1.73 billion and $2.34 billion and the BLM backlog at between $0.39 billion and $0.47 billion. The four agencies together have a combined backlog estimated at between $13.88 billion and $21.20 billion, with a mid-range figure of $17.54 billion, according to the agencies [4]. The NPS and other agency backlogs have been attributed to decades of funding shortfalls. The agencies assert that continuing to defer maintenance of facilities accelerates their rate of deterioration, increases their repair costs, and decreases their value.

Administrative Actions. In FY2002, the Bush Administration proposed to eliminate the NPS backlog (estimated at $4.9 billion in 2002) over five years. The NPS budget justification for FY2007 states that, beginning with FY2002, "nearly $4.7 billion has been invested in deferred maintenance." [5]. The figure reflects total appropriations for line items of which deferred maintenance is only a part. Specifically, according to the NPS, it consists of appropriations for all NPS facility maintenance, NPS construction, and the NPS park roads and parkway program funded through the Federal Highway Administration. It also includes fees used for maintenance. The National Parks Conservation Association claims that the Administration has supported little new money to address park maintenance, and is using "misleading" math to appear to be on track to eliminate the backlog [6]. It further contends that national parks on average have about 2/3 of the funding they need, and that sufficient operating funds are necessary for stemming the growth of the backlog.

It is uncertain if the NPS backlog has decreased, increased, or remained the same in recent years. For instance, while estimates of the backlog increased from an average of $4.25 billion in FY1999 to $9.11 billion in FY2005, it is unclear what portion of the change is due to the addition of maintenance work that was not done on time or the availability of more precise estimates of the backlog. Further, it is unclear how much total funding has been provided for backlogged maintenance over this period. Annual presidential budget requests and appropriations laws do not typically specify funds for backlogged maintenance, but instead combine funding for all NPS construction, facility operation, and regular and deferred maintenance. According to the DOI Budget Office, the appropriation for NPS deferred maintenance increased from $223.0 million in FY1999 to $311.1 million in FY2006, with a peak in FY2002 at $364.2 million [7]. For FY2007, the Administration requested $208.1 million, a $103.0 million (33%) reduction from the FY2006 level and a $14.9 million (7%) reduction from the FY1999 level.

The NPS has been defining and quantifying its maintenance needs. These efforts, like those of other land management agencies, include developing computerized systems for tracking and prioritizing maintenance projects and collecting comprehensive data on the condition of facilities — expected by the end of FY2006.

Legislative Activity. H.R. 1124 and S. 886 seek to eliminate the annual operating deficit and maintenance backlog in the National Park System. They would create the National Park Centennial Fund in the Treasury, to be comprised of monies designated by taxpayers on their tax returns. If monies from tax returns are insufficient to meet funding levels established in the bill, they are to be supplemented by contributions to the Centennial Fund from the General Fund of the Treasury. For FY2006, there is to be deposited in the Centennial Fund $150.0 million, with an increase of 15% each year though FY2016. The Fund would be available to the Secretary of the Interior, without further appropriation, as follows: 60% to eliminate the NPS maintenance backlog, 20% to protect NPS natural resources, and 20% to protect NPS cultural resources. The Senate bill would terminate the fund on October 1, 2016. Under the House bill, after that date money in the Centennial Fund is to be used to supplement annual appropriations for park operations. The bills also would require the Government Accountability Office (GAO) to submit to Congress biennial reports on the progress of the NPS in eliminating its deficit in operating funds and the funding needs of national parks compared with park appropriations, among other issues. In addition, on May 10, 2005, a Senate subcommittee held a hearing on NPS funding issues, including the maintenance backlog.

Policy Revisions

Background. On August 31, 2006, the NPS issued new service-wide management policies that govern the way NPS managers make decisions on a wide range of issues (together with laws, regulations, and other authorities) [8]. The final policies appear to have met with broad support from earlier critics of an October 19, 2005 draft, although some recreation advocates prefer the emphasis of the draft over the final version. The process by which the policies were reformed was controversial. The NPS Management Policies were last updated in 2001 after a several-year internal and external review.

Administrative Actions. The NPS, in a news release, cited varied reasons for initiating a revision of its management policies, namely legal, social, and technological developments since the issuance of the 2001 policies. With regard to legal issues, the agency notes new laws, executive orders, and regulations affecting park management. Social issues appear to include increased responsibilities for homeland security, population growth near parks, changes in park visitation, and a focus on strengthening community ties. Technological issues stem from developments that provide new ways to recreate in parks or reduce adverse affects on resources [9]. According to an NPS spokesman, policy revisions also were undertaken because coverage of financial issues was needed, including on recreation fees, concession royalties, and Park Service donations, and because there was some support in Congress for a review of NPS management policies.

The development of policy changes began with the preparation of draft changes by a senior DOI official. That initial, internal proposal was intended to promote in-house discussion of management policies, according to DOI. Nevertheless, it was criticized by some park groups and environmentalists as shifting the NPS focus from preservation to recreation; removing protective limits on activities that might impair park resources, for instance, motorized recreation; eliminating the scientific underpinning of NPS management; giving too much control to local communities in managing park units; weakening protections for air quality, water, and wildlife; and increasing commercial development of park units. Further, some observers criticized DOI for initiating changes to NPS policies without notifying NPS employees and consulting with the public. That initial draft was reported by the press to have been opposed by the NPS's seven regional directors. The agency subsequently convened a working group of 16 senior staff, who produced a new draft. That draft was to have been reviewed, including by the National Leadership Council — a group of senior park managers who set policy and overall direction for the NPS, before its publication.

On October 19, 2005, the NPS published draft Management Policies (70 *Fed. Reg.* 60852), with a public comment period through February 18, 2006. Some park groups and environmentalists were concerned that changes would fundamentally alter park protections and potentially lead to damaging park resources. One much-discussed proposal sought to require "balance" between conservation and enjoyment of park resources, whereas the then-existing policy stated that "conservation is to be predominant" in conservation/enjoyment conflicts. This controversy illustrates a long-standing tension in the Park Service's mission to protect park resources while providing for their use and enjoyment by the public.

The NPS received approximately 45,000 comments, and made revisions to the draft policies based on these comments. The draft underwent further review, for example by the National Leadership Council. On June 19, 2006, the NPS issued revised draft management polices. That version was widely viewed as shifting park priorities back to preservation, and was thus generally supported by conservation interests. Some critics viewed the policies as favoring conservation over recreation, and thus as insufficiently allowing for public use and enjoyment of NPS lands and resources. Others viewed the policies as failing to address or resolve certain issues. After a final review and relatively minor revisions, the polices were made final on August 31, 2006.

The final policies contain a list of underlying principles, including that the policies must "ensure that conservation will be predominant when there is a conflict between the protection of resources and their use" (p. iv). In testimony on June 20, 2006, the NPS Deputy Director outlined the "improvement" from the 2001 to 2006 policies, primarily changes in emphasis and clarity in many areas. They include a commitment to civic engagement, cooperative conservation, and improvements in workforce and business practices. Other changes involve additional guidance on relationships between parks and Native Americans, and recognition of the importance of clean air, clean water, and soundscapes. Still other changes involve new guidance on determining what is an appropriate or inappropriate use of parks, and management of uses to avoid impairment of resources [10]. In testimony on July 25, 2006, the NPS Director further elaborated that the 2006 policies ensure that Americans will continue to enjoy national parks.

Legislative Activity. House and Senate committees have held several hearings on park management policies. For instance, on November 1, 2005, a Senate subcommittee held a hearing on the draft policies. Witnesses expressed differing opinions on issues including the reasons the policies are being revised; the intent of the 1916 Park Service Organic Act regarding preservation and recreation; the extent to which the policies should emphasize recreation; the impact of proposed changes on park protections and the impairment standard; and whether the draft changes would blur or clarify how park employees are to manage resources.

The NPS Organic Act and its implementation through daily park management were the subject of a December 14, 2005, House Resources subcommittee hearing. Witnesses offered different views on the intent of the NPS Organic Act, particularly with regard to preservation, use, and impairment of NPS resources. Witnesses also presented varying opinions on whether then-existing park management policies, or the proposed policy revisions, more accurately reflect the letter and intent of the Organic Act. Whether the management policies should be rewritten, and the proposed changes themselves, also were a matter of much debate. Some witnesses claimed that the NPS has limited access for recreation in recent years, in favor of preservation of resources, and suggested alternative approaches. In addition, at a February 15, 2006, hearing, the subcommittee heard differing views from Administration and private witnesses as to whether park policies should be changed and whether the particular changes in the draft would be beneficial or detrimental.

A June 20, 2006, Senate Energy and Natural Resources subcommittee took testimony from two witnesses on the revisions to park management policies, primarily the June 19, 2006, version. The NPS Deputy Director described the process for revising the policies, and differences between the 2001 and June 19, 2006, park policies (discussed above under "Administrative Actions"). A witness from the National Parks Conservation Association expressed general support for the June 19, 2006, version of the park policies, as apparently having "discarded the broad changes that caused so much national concern" [11]. His testimony compared selected provisions of the 2001 policies with the June 2006 and earlier proposed revisions.

A subcommittee of the House Resources Committee held a hearing on park management policies on July 25, 2006. In an opening statement, the subcommittee chairman expressed the concern of some Members that the June 19, 2006, policy proposals subordinate the enjoyment and use of park units to the conservation of resources. He expressed support for the October 2005 proposals as more accurately reflecting the mission of the NPS, and compared relevant provisions of that draft with the June 19, 2006, one. Further, the chairman outlined "failures" in the park system that were not addressed by the latest draft policies, including inconsistency in implementing construction requirements under the Americans with Disabilities Act and actions that the NPS should take to increase visitation [12]. As the sole witness, the NPS Director provided an overview of the areas of emphasis of the June 19, 2006, draft policies, and also described the changes as an "improvement" over the 2001 policies. She summarized the process through which the NPS was developing revised policies [13].

Congressional hearings have been held on park management issues other than the policy revisions. For instance, with regard to park management generally, a subcommittee of the House Government Reform Committee is in the midst of a series of oversight hearings on the role and management of park units. These hearings, being held throughout the country, are examining the issues facing the variety of park units in different areas of the country. They have encompassed diverse issues, including the adequacy of park budgets, backlog in maintaining NPS facilities, control of invasive species, nature and extent of visitor services, and protection of park resources. A report summarizing the critical issues discussed, together with recommendations, is anticipated at the conclusion of the hearings.

Wild and Scenic Rivers

Background. The NPS manages 28 river units, totaling 2,826.3 miles, within the National Wild and Scenic Rivers System. The system was authorized on October 2, 1968, by the Wild and Scenic Rivers Act (16 U.S.C. §§1271-1287) [14]. The act established a policy of preserving designated free-flowing rivers for the benefit and enjoyment of present and future generations, to complement the then-current national policy of constructing dams and other structures along many rivers. The act requires that river units be classified and administered as wild, scenic, or recreational rivers, based on the condition of the river, the amount of development in the river or on the shorelines, and the degree of accessibility by road or trail at the time of designation.

Typically rivers are added to the system by an act of Congress, but they also may be added by state nomination with the approval of the Secretary of the Interior. Congress initially designated 789 miles of 8 rivers as part of the system. Today there are 164 river units with 11,357.7 miles in 38 states and Puerto Rico, administered by the NPS, other federal agencies, and several state agencies. Congress also commonly enacts legislation to authorize the study of particular rivers for potential inclusion in the system. The NPS maintains a national registry of rivers that may be eligible for inclusion in the system — the Nationwide Rivers Inventory (NRI) [15]. Congress may consider, among other sources, these NRI rivers which are believed to possess "outstandingly remarkable" values. The Secretaries of the Interior and Agriculture are to report to the President as to the suitability of study areas for wild and scenic designation. The President then submits his recommendations regarding designation to Congress.

Administrative Actions. Wild and scenic rivers designated by Congress generally are managed by one of the four federal land management agencies — NPS, FWS, BLM, and FS. Management varies with the class of the designated river and the values for which it was included in the system. Components of the system managed by the NPS become a part of the National Park System. The act requires the managing agency of each component of the system to prepare a comprehensive management plan to protect river values. The managing agency also establishes boundaries for each component of the system, within limitations. Management of lands within river corridors has been controversial in some cases, with debates over the effect of designation on private lands within the river corridors, the impact of activities within a corridor on the flow or character of the designated river segment, and the extent of local input in developing management plans.

State-nominated rivers may be added to the National Wild and Scenic Rivers System only if the river is designated for protection under state law, is approved by the Secretary of the Interior, and is permanently administered by a state agency. Management of state-nominated rivers may be complicated because of the diversity of land ownership.

Legislative Activity. Measures to designate, study, or extend specific components of the Wild and Scenic Rivers System are shown in the following table. The table includes bills that could involve management by the NPS or other agencies.

Table

Title	Type		Status
Alaska Rainforest Conservation Act (designate and study rivers within the Chugach NF and designate rivers within the Tongass NF)	Desig./ Study	H.R. 1155	Introduced
California Wild Heritage Act of 2006 (designate 22 river segments; study Carson River, East Fork, CA)	Desig./ Study	H.R. 5006 S. 2432	Introduced Introduced
Eastern Sierra Rural Heritage and Economic Enhancement Act (designate segments of the Amargosa River in CA)	Desig.	H.R. 5149 S. 2567	Hearing Held Hearing Held
Eightmile Wild and Scenic River Act (CT)	Desig.	H.R. 5885 S. 3723	Introduced Introduced
Fossil Creek Wild and Scenic River Act (AZ)	Desig.	H.R. 5957 S. 3762	Introduced Introduced
Lower Farmington River and Salmon Brook Wild and Scenic River Study Act of 2005 (CT)	Study	H.R. 1344 S. 435	Passed House Passed Senate
Mt. Hood Stewardship Legacy Act Lewis and Clark Mount Hood Wilderness Act (both to designate waterways in the Mt. Hood NF (OR))	Desig.	H.R. 5025 S. 3854	Passed House Introduced

Table (Continued)

Title	Type		Status
Musconetcong Wild and Scenic Rivers Act (NJ)	Desig.	H.R. 1307 S. 1096	Passed House Passed Senate
Northern California Coastal Wild Heritage Wilderness Act (Black Butte River segments)	Desig.	H.R. 233 S. 128	Senate Calendar Passed Senate
Owyhee Initiative Implementation Act (ID) (to designate rivers in Idaho)	Desig.	S. 3794	Introduced
Perquimans River Wild and Scenic River Study Act (NC)	Study	H.R. 4105	Introduced
Rockies Prosperity Act (Title IV, to designate certain National Forest System watercourses in ID, MT, and WY)	Desig.	H.R. 1204	Introduced
Taunton Wild and Scenic Rivers Act (MA)	Desig.	H.R. 3321 S. 2033	Introduced Introduced
Upper White Salmon Wild and Scenic Rivers Act (WA)	Desig.	H.R. 38 S. 74	P.L. 109-44; Indef. Postponed
Washington County Growth and Conservation Act (to designate segments of the Virgin River and its tributaries across federal land within and adjacent to Zion National Park, UT)	Desig.	S. 3636 H.R. 5769	Introduced Hearing Held

Chamber-Passed Bills. The 109th Congress has enacted H.R. 38 (P.L 109-44) to designate a portion of the White Salmon River (WA) as a component of the National Wild and Scenic Rivers System. Several other bills have passed the House or Senate, as discussed below.

On December 16, 2005, the Senate passed S. 435, to direct the NPS to study a 40-mile stretch of the Farmington River and Salmon Brook (CT) for possible inclusion in the National Wild and Scenic Rivers System. As a result of reduced funding for the Rivers and Trails Studies program for FY2006, the NPS had requested that the date for submitting the study be changed from not later than three years following enactment to not later than three years after funds are made available. This change is included in the Senate-passed bill. Some river proponents objected to the delay in the start of the study. A companion bill, H.R. 1344, passed the House on September 25, 2006.

Also on December 16, 2005, the Senate passed S. 1096 — the Musconetcong Wild and Scenic Rivers Act — to designate 24.2 miles of the river in northwestern New Jersey. The House passed a companion bill, H.R. 1307, on July 24, 2006. Both bills also provide for the designation of an additional 4.3 miles of the Musconetcong River as a recreational river, if the Secretary of the Interior determines that there is adequate local support.

On July 24, 2006, the House also passed H.R. 5025, to designate 25 miles of waterways in the Mount Hood National Forest (OR) as additions to the Wild and Scenic Rivers System. A Senate companion bill, S. 3854, seeks to designate the same segments and additional ones for a total of 81 miles of waterways in the Mount Hood National Forest. The bill has been referred to committee.

Both the House and Senate have passed companion legislation, H.R. 233 and S. 128, to designate segments totaling 21 miles of the Black Butte River (CA) as a wild or scenic river. Both bills also require the Secretary of Agriculture to report to Congress regarding a fire management plan for the segments and the cultural and historic resources in those segments.

OTHER ISSUES

Competitive Sourcing. (by Carol Hardy Vincent) The Bush Administration's Competitive Sourcing Initiative seeks to expand on earlier programs to subject federal agency activities judged to be commercial in nature to public-private competition. The Administration's goal is to save money through competition. For the NPS, areas of focus include maintenance, administration, and cultural resource positions. Rangers, fee collectors, and park guides are among those positions classified as either "inherently governmental" or "core to the mission," and thus not subject to competitive review. Concerns include whether the initiative would save the agency money, whether it is being used to accomplish policy objectives by outsourcing particular functions, whether it would weaken the morale and diversity of the NPS workforce, and whether the private sector could provide the same quality of service. The NPS has long contracted many jobs to private industry. (For information on competitive sourcing generally, see CRS Report RL32017, *Circular A-76 Revision 2003: Selected Issues*, by L. Elaine Halchin).

The NPS competitive sourcing "green plan" covers competitive sourcing activities planned for FY2005-FY2008. During FY2006, the NPS planned to conduct a preliminary planning effort for 150 FTEs, [16] four standard studies for 549.5 FTEs, and six streamlined studies for 255.5 FTEs, for a total of 955 FTEs during FY2006. For FY2007, the agency expects to review about 700 FTEs and subsequently to implement related efficiencies. (For information on competitive sourcing targets, see CRS Report RL32079, *Federal Contracting of Commercial Activities: Competitive Sourcing Targets*, by L. Elaine Halchin).

P.L. 109-54, the FY2006 Interior, Environment, and Related Agencies Appropriations Act, placed a cap of $3.45 million on DOI competitive sourcing studies during FY2006, but did not specify the portion to be allocated to the NPS. The law also provided that agencies include, in any reports to the Appropriations Committees on competitive sourcing, information on costs associated with sourcing studies and related activities. The House, and the Senate Committee on Appropriations, included similar provisions for FY2007 in their versions of H.R. 5386, the Interior, Environment, and Related Agencies Appropriations bill for FY2007. These provisions originated out of concern that some agencies were spending significant sums on competitive sourcing where the Administration did not request or receive funds for this purpose, and were not providing Congress with complete information on costs and implications. P.L. 109-115 restricts competitive sourcing government-wide.

Regional Haze. (by Ross W. Gorte) In 1977 amendments to the Clean Air Act, Congress established a national goal of protecting Class I areas — most then-existing national parks and wilderness areas — from future visibility impairment and remedying any existing impairment resulting from manmade air pollution. (Newly designated parks and wilderness areas can be classified as Class I only by state actions.) The program to control this "regional haze" has several facets, including the development of state implementation plans and the imposition of Best Available Retrofit Technology (BART) on large sources of air pollution built between 1962 and 1977. (For a general description of the regional haze program, see CRS Report RL32483, *Visibility, Regional Haze, and the Clean Air Act: Status of Implementation,* by Larry Parker and John Blodgett). A related program, Prevention of Significant Deterioration, provides that permits may not be issued to major new facilities within 100 kilometers of a Class I area if federal land managers, such as at the NPS, allege that the facilities' emissions "may cause or contribute to a change in the air quality" in a Class I area (42 U.S.C. §7457).

DOI's strategic plan (2004) contains two air quality goals for Class I areas, related to compliance with national ambient air quality standards and visibility objectives. At 68 park units, the NPS monitors one or more key air quality indicators, such as ozone, visibility, and atmospheric deposition, and reports annually on progress towards meeting air quality goals. The latest report (2005) examined data collected between 1995-2004. It concluded that of the reporting park units, 68% showed stable or improving air quality trends generally, 78% met national ambient air quality standards, and 100% met visibility goals. The agency expressed that meeting air quality goals is challenging because the NPS does not have direct authority to control pollution sources outside of park units. Nevertheless, NPS expects further improvement in meeting goals as regulations to reduce tailpipe emissions from motor vehicles and pollution from electric-generating facilities take full effect [17]. In August 2006, the National Parks Conservation Association released a new report asserting that "air pollution is among the most serious and wide-ranging problems facing the parks today.... We've made some important advances ... but much more remains to be done." [18]. The report includes 10 recommendations to improve air quality in the National Park System.

Security. (by Carol Hardy Vincent) Since the September 11, 2001 terrorist attacks on the United States, the NPS has sought to enhance its ability to prepare for and respond to threats from terrorists and others. Activities have focused on security enhancements at national icons and along the U.S. borders, where several parks are located. The United States Park Police (USPP) have sought to expand physical security assessments of monuments, memorials, and other facilities, and increase patrols and security precautions in Washington monumental areas, at the Statue of Liberty, and at other potentially vulnerable icons. Other activities have included implementing additional training in terrorism response for agency personnel, and reducing the backlog of needed specialized equipment and vehicles. NPS law enforcement rangers and special agents have expanded patrols, use of electronic monitoring equipment, intelligence monitoring, and training in preemptive and response measures. The NPS has taken measures to increase security and protection along international borders and to curb illegal immigration and drug traffic through park borders.

A June, 2005 report of the Government Accountability Office (GAO) examined the challenges for DOI in protecting national icons and monuments from terrorism, and actions and improvements the department has taken in response [19]. GAO concluded that since 2001, DOI has improved security at key sites, created a central security office to coordinate security efforts, developed physical security plans, and established a uniform risk management and ranking methodology. GAO recommended that DOI link its rankings to security funding priorities at national icons and monuments and establish guiding principles to balance its core mission with security needs. (See [http://www.gao.gov/new.items/ d05790.pdf]).

Several 109th Congress hearings have been held on illegal border issues affecting federal lands along the northern and southern U.S. borders, including NPS lands. Hearings have addressed the adverse affects of such activities on federal lands, how to reduce harm from illegal border activities, efforts of various agencies to secure federal lands along the borders, and the demands on law enforcement personnel of the federal land management agencies. Illegal activities at issue have included drug trafficking, alien smuggling, money laundering, organized crime, and terrorism. Such activities are reported to have caused damage to federal lands, including by creating illegal roads, depositing large amounts of trash and human waste, increasing risk of fire from poorly tended camp fires, destroying vegetation and cultural resources, and polluting waterways. The effects on federal lands of border enforcement activities, in response to illegal immigration, also has been addressed. Some agency witnesses discussed the implementation of a recent memorandum of understanding between the Departments of Homeland Security, Interior, and Agriculture on initiatives to improve handling of illegal border activities and their impacts on federal lands.

House and Senate bills pertaining to immigration reform and border security contain provisions affecting national park units along U.S. borders. For example, as passed by the House, H.R. 4437 would require an evaluation of security vulnerabilities on DOI lands along U.S. borders and would require the Secretary of Homeland Security to provide border security assistance on these lands. S. 2611, which has passed the Senate, calls for a study of the construction of physical barriers along the southern border of the United States, including their effect on park units along the borders. Among other provisions, S. 2611 also would increase customs and border protection personnel to secure park units (and other federal land) along U.S. borders; provide surveillance camera systems, sensors, and other equipment for lands on the border, with priority for NPS units; and require a recommendation to Congress for the NPS and other agencies to recover costs related to illegal border activity.

Congress appropriates funds to the NPS for security efforts, and the adequacy and use of funds to protect NPS visitors and units are of continuing interest. Funds for security are appropriated through multiple line items, including those for the USPP and Law Enforcement and Protection. For FY2007, the President requested $84.8 million for the USPP, a 6% increase over FY2006 ($80.2 million). The House and the Senate Committee on Appropriations approved this level in H.R. 5386, the FY2007 Interior, Environment, and Related Agencies appropriations bill. The President also requested $128.2 million for law enforcement, a 3% increase over FY2006 ($124.2 million). The amount approved by the House or the Senate Committee on Appropriations was not specified in H.R. 5386.

FOR ADDITIONAL READING

CRS Report RL32017, *Circular A-76 Revision 2003: Selected Issues*, by L. Elaine Halchin.

CRS Report RL32833, *Competitive Sourcing Legislation*, by L. Elaine Halchin.

CRS Report RL32079, *Federal Contracting of Commercial Activities: Competitive Sourcing Targets*, by L. Elaine Halchin.

CRS Report RL32393, *Federal Land Management Agencies: Background on Land and Resources Management*, coordinated by Carol Hardy Vincent.

CRS Report RL33596, *Federal Lands Managed by the Bureau of Land Management (BLM) and the Forest Service*, coordinated by Ross W. Gorte and Carol Hardy Vincent.

CRS Report RL32667, *Federal Management and Protection of Paleontological (Fossil) Resources Located on Federal Lands: Current Status and Legal Issues*, by Douglas Reid Weimer.

CRS Report RL33462, *Heritage Areas: Background, Proposals, and Current Issues*, by Carol Hardy Vincent and David Whiteman.

CRS Report RL33617, *Historic Preservation: Background and Funding*, by Susan Boren.

CRS Report RS22298, *Historic Preservation: Federal Laws and Regulations Related to Hurricane Recovery and Reconstruction*, by Douglas Reid Weimer.

CRS Report RL33399, *Interior, Environment, and Related Agencies: FY2007 Appropriations*, coordinated by Carol Hardy Vincent and Susan Boren.

CRS Report RL33531, *Land and Water Conservation Fund: Overview, Funding History, and Current Issues*, by Carol Hardy Vincent.

CRS Report RS20902, *National Monument Issues*, by Carol Hardy Vincent.

CRS Report RS20158, *National Park System: Establishing New Units*, by Carol Hardy Vincent.

CRS Report RL32699, *Natural Resources: Selected Issues for the 109th Congress*, coordinated by Carol Hardy Vincent, Nicole T. Carter, and Julie Jennings.

CRS Report RL33525, *Recreation on Federal Lands*, coordinated by Kori Calvert and Carol Hardy Vincent.

CRS Report RL31149, *Snowmobiles: Environmental Standards and Access to National Parks*, by James E. McCarthy.

CRS Report RS20702, *South Florida Ecosystem Restoration and the Comprehensive Everglades Restoration Plan*, by Pervaze A. Sheikh and Nicole T. Carter.

CRS Report RL32483, *Visibility, Regional Haze, and the Clean Air Act: Status of Implementation*, by Larry Parker and John Blodgett.

CRS Report RL30809, *Wild and Scenic Rivers Act and Federal Water Rights*, by Pamela Baldwin.

CRS Report RL31447, *Wilderness: Overview and Statistics*, by Ross W. Gorte.

REFERENCES

[1] Descriptions of the different designations are on the NPS website at [http://www.nps.gov/legacy/]. Brief information on each unit is contained in U.S. Dept. of the

Interior, National Park Service, *The National Parks: Index 2001-2003* (Washington, DC: 2001).

[2] This figure includes an estimated 79 million acres of federal land, 1 million acres of other public land, and 4 million acres of private land within unit boundaries. NPS policy is to acquire these nonfederal *inholdings* from willing sellers as funds are made available or to create special agreements to encourage landowners to sell.

[3] *Federal Historic Rehabilitation Tax Credit Program: Recommendations for Making a Good Program Better*, Report to the National Park System Advisory Board by the Committee on the Federal Historic Rehabilitation Tax Credit Program. September, 2006.

[4] Estimates are from DOI and the FS, and reflect only direct project costs in accordance with requirements of the Federal Accounting Standards Advisory Board.

[5] U.S. Dept. of the Interior, National Park Service, *Budget Justifications and Performance Information, Fiscal Year 2007*, p. overview-3 (Washington, DC: 2006).

[6] National Parks Conservation Association, *The Burgeoning Backlog: A Report on the Maintenance Backlog in America's National Parks* (May 2004), p. 6, available on the web at [http://www.npca.org/across_the_nation/visitor_experience/backlog/ backlog.pdf]

[7] U.S. Dept. of the Interior, Office of Budget, Internal Memorandum, Washington, D.C., received April 7, 2006.

[8] The new policies are available on the NPS website at [http://home.nps.gov/applications/ npspolicy/index.cfm].

[9] National Park Service, news release of June 19, 2006, entitled *Kempthorne: Park Management Policies Will Assure Legacy of Conservation*, available on the NPS website at [http://home.nps.gov/applications/release/Detail.cfm?ID=681].

[10] National Park Service, testimony of Stephen P. Martin, June 20, 2006, available on the website of the Senate Committee on Energy and Natural Resources at [http://energy.senate.   gov/public/index.cfm?FuseAction=Hearings.Testimonyand Hearing_ID=1564and Witness_ ID=4235].

[11] National Parks Conservation Association, testimony of Thomas C. Kiernan, June 20, 2006, available on the website of the Senate Committee on Energy and Natural Resources at: [http://energy.senate.gov/public/index.cfm?FuseAction=Hearings. Testimonyand Hearing _ID=1564and Witness_ID=2337].

[12] Opening Statement of Chairman Stevan Pearce, Subcommittee on National Parks, Committee on Resources, Oversight Hearing on NPS Management Polices, July 25, 2006.

[13] National Park Service, testimony of Fran Mainella, July 25, 2006, available on the website of the House Committee on Resources at [http://resourcescommittee.house.gov/ archives/ 109/nprpl/072506.htm].

[14] The text of the Wild and Scenic Rivers Act is available on the NPS website at [http://www.nps.gov/rivers/wsract.html].

[15] For further Information on the Nationwide Rivers Inventory, see the NPS website at [http://www.nps.gov/rtca/nri/].

[16] A full-time equivalent (FTE) is the "staffing of Federal civilian employee positions, expressed in terms of annual productive work hours" (U.S. Office of Management and Budget, *Circular No. A-76 (Revised)*, p. D-5).

[17] See *2005 Annual Performance and Progress Report: Air Quality in National Parks*, available on the NPS website at [http://www2.nature.nps.gov/air/Pubs/index.cfm].

[18] National Parks Conservation Association, *Turning Point*, p. 4, available on the web at [http://www.npca.org/turningpoint/Full-Report.pdf].

[19] Government Accountability Office, *Homeland Security: Actions Needed to Better Protect National Icons and Federal Office Buildings from Terrorism*, GAO-05-790 (DC: June 2005).

In: Environmental Research Advances
Editor: Peter A. Clarkson, pp. 135-154

ISBN: 978-1-60021-762-3
© 2007 Nova Science Publishers, Inc.

Chapter 6

ENVIRONMENTAL INFORMATION EPA ACTIONS COULD REDUCE THE AVAILABILITY OF ENVIRONMENTAL INFORMATION TO THE PUBLIC*

John B. Stephenson

ABSTRACT

Although we have not yet completed our review, our preliminary observations are that EPA did not adhere to all aspects of its rulemaking guidelines when developing the new TRI reporting requirements. EPA's Action Development Process outlines a series of steps to help guide the development of new environmental regulations. Throughout this process, however, the senior EPA management has the authority to accelerate the rule development process. Nevertheless, while we continue to pursue a clearer understanding of EPA's actions, we have identified several significant differences between the guidelines and the process EPA followed in this case: (1) late in the rulemaking process, senior EPA management directed consideration of a burden reduction option that the TRI workgroup had previously dropped from consideration; (2) EPA developed this option on an expedited schedule that appears to have provided a limited amount of time for conducting various impact analyses; and (3) EPA's decision to expedite Final Agency Review, when EPA's internal and regional offices determine whether they concur with the final proposal appears to have limited the amount of input they could provide to senior EPA management. First, the TRI workgroup charged with identifying options to reduce reporting burdens on industry identified three possible options for senior management to consider. The first two options allowed facilities to use Form A in lieu of Form R for PBT chemicals, provided the facility has no releases to the environment, and the third created a "no significant change" reporting option in lieu of Form R for facilities with releases that changed little from the previous year. Information from a June 2005 briefing for the Administrator indicated that, while the Office of Management and Budget (OMB) had suggested increasing the Form A eligibility for non-PBT chemicals from 500 to 5,000 pounds, the TRI workgroup dropped that option from consideration. Second, although we could not determine from the documents provided by EPA what actions the agency took between the June 2005 briefing for the EPA Administrator and the October

---

* Excerpted from GAO 07-464T, dated February 6, 2007.

2005 issuance of the TRI proposal in the *Federal Register*, the Administrator provided direction after the briefing to expedite the process in order to meet a commitment to OMB to provide burden reduction by the end of December 2006. Subsequently, EPA revised its economic analysis to include consideration of the impact of raising the Form A eligibility threshold. However, that analysis was not completed before EPA sent the proposed rule to OMB for review and was only completed just prior to the proposal being signed by the Administrator and published in the *Federal Register* for public comment.

Third, the extent to which senior EPA management sought or received input from internal stakeholders, including the TRI workgroup, after directing reconsideration of the option to increase the Form A reporting threshold from 500 to 5,000 pounds for non-PBT chemicals remains unclear. We have been unable to determine the extent to which EPA's internal and regional offices had the opportunity during Final Agency Review to determine whether they concurred with the proposal to increase the Form A threshold. We will continue to pursue the answer to this and other questions as we complete our work. Finally, in response to the public comments on the proposal, nearly all of which were negative, EPA considered alternative options and revised the proposal, thereby allowing facilities to report releases of up to 2,000 rather than 5,000 pounds on Form A.

We believe that the TRI reporting changes will likely have a significant impact on information available to the public about dozens of toxic chemicals from thousands of facilities in states and communities across the country. EPA estimated that the TRI reporting changes will affect reporting on less than 1 percent of the total chemical releases reported to the TRI annually. While our analysis supports EPA's estimate of this impact at a national level, it also suggests that changes to TRI reporting requirements will have a significant impact on the amount and nature of toxic release data available to some communities, information that is ultimately much more meaningful to citizens. In addition, preliminary results from our January 2007 survey of state TRI coordinators indicates that as many as 23 states believe that EPA's changes to TRI reporting requirements will have a negative impact on various aspects of TRI. To develop a more specific picture of the impact of the TRI reporting changes at a local level, we used 2005 TRI data to estimate, by state, the impact of EPA's changes. First, we estimated that the detailed information from more than 22,000 Form R reports may no longer be included in the TRI if all eligible facilities begin using Form A. More specifically, Alaska, California, Connecticut, Hawaii, Massachusetts, New Jersey, and Rhode Island could have 33 percent fewer chemical reports. Second, we estimated that the number of chemicals for which no information could be reported under the new rule ranges from 3 chemicals in South Dakota to 60 chemicals in Georgia. Thirteen states—including Delaware, Georgia, Maryland, Missouri, Oklahoma, Tennessee, and Vermont—could have no detailed reports on more than 20 percent of reported chemicals. Third, we estimated that a total of 3,565 facilities would no longer have to report quantitative information about their chemical use to the TRI. In fact, more than 20 percent of facilities in Colorado, Connecticut, Hawaii, Massachusetts, and Rhode Island, could have no detailed information about their chemical use. Furthermore, citizens living in 75 counties in the United States—including 11 in Texas, 10 in Virginia, and 6 in Georgia—could have no quantitative TRI information about local toxic pollution. Finally, with regard to the impact of the rule change on industry's reporting burden, EPA estimated that, if all eligible facilities take advantage of the reporting changes, they could save a total of about $5.9 million—about 4 percent of the annual cost of TRI reporting. This is the equivalent of less than $900 per facility. However, based on past experience, not all eligible facilities will use Form A, so the actual savings to industry are likely to be less.

With regard to your request for an update on our May 2005 report on perchlorate, it should be noted that perchlorate releases are not reported to the TRI. Perchlorate, a primary ingredient in propellant, has been used for decades in the manufacture and firing of rockets and missiles. Other uses include fireworks, flares, and explosives. Perchlorate is a salt that is easily dissolved and transported in water and has been found in groundwater, surface water, drinking water, soil, and food products such as milk and lettuce across the country. Health studies have shown that perchlorate can affect the thyroid gland and may cause developmental delays. We identified more than 400 sites in 35 states where perchlorate had been found in concentrations ranging from 4 parts per billion to more that 3.7 million parts per billion, and that more than one-half of the sites were in California and Texas. However, federal and state agencies are not required to routinely report perchlorate findings to EPA, and EPA does not centrally track or monitor perchlorate detections or the status of cleanup efforts. As a result, a greater number of contaminated sites than we reported may exist. Although concern over potential health risks from perchlorate has increased, and at least 9 states have established nonregulatory action levels or advisories, EPA has not established a national drinking water standard citing the need for more research on health effects. We concluded in our report that EPA needed more reliable information on the extent of sites contaminated with perchlorate and the status of cleanup efforts, and recommended that EPA work with the Department of Defense and the states to establish a formal structure for tracking perchlorate information. In December 2006, EPA reiterated its disagreement with the recommendation stating that perchlorate information already exists from a variety of other sources. However, we continue to believe that the inconsistency and omissions in available data that we found during the course of our study underscore the need for a more structured and formal tracking system.

BACKGROUND

In 1984, a catastrophic accident caused the release of methyl isocyanate—a toxic chemical used to make pesticides—at a Union Carbide plant in Bhopal, India, killing thousands of people, injuring many others, and displacing many more from their homes and businesses. One month later, it was disclosed that the same chemical had leaked at least 28 times from a similar Union Carbide facility in Institute, West Virginia. Eight months later, 3,800 pounds of chemicals again leaked from the West Virginia facility, sending dozens of injured people to local hospitals. In the wake of these events, Congress passed the Emergency Planning and Community Right-to-Know Act of 1986 (EPCRA). Among other things, EPCRA provides access by individuals and communities to information regarding hazardous materials in their communities. Section 313 of EPCRA generally requires certain facilities that manufacture, process, or otherwise use any of 581 individual chemicals and 30 additional chemical categories to annually report the amount of those chemicals that they released to the environment, including information about where they released those chemicals. EPCRA also requires EPA to make this information available to the public, which the agency does in a national database known as the Toxics Release Inventory. The public may access TRI data on EPA's website and aggregate it by zip code, county, state, industry, and chemical. EPA also publishes an annual report that summarizes national, state, and industry data [1].

Figure 1 illustrates TRI reporting using a typical, large coal-fired electric power plant as an example [2]. The figure notes the chemicals that the facility may have to report to the TRI. The primary input to this facility is coal that contains small amounts of a number of toxic chemicals such as arsenic, chromium, and lead. The facility pulverizes coal and burns it to generate electricity. As part of its standard operations, the facility releases TRI chemicals such as hydrochloric acid and sulfuric acid to the air through its stack. The facility may also send ash from the burning process to an ash pond or landfill, including TRI chemicals such as arsenic, lead, and zinc. In addition, the facility may release chemicals in the water it uses for cooling. The facility will have to complete a TRI report for air, land, and water releases of each chemical it uses above a certain threshold.

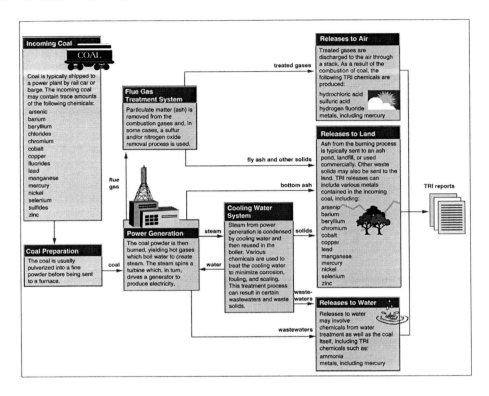

Figure 1. TRI Reporting at a Typical Coal-fired Electric Generation Facility.

Owners of facilities subject to EPCRA comply its reporting requirements by submitting an annual Form R report to EPA, and their respective state, for each TRI-listed chemical that they release in excess of certain thresholds. Form R captures information about facility identity, such as address, parent company, industry type, latitude, and longitude and detailed information about the toxic chemical, such as quantity of the chemical disposed or released onsite to air, water, land, and underground injection or transferred for disposal or release off-site. This information is labeled as "Disposal or Other Releases" on the left side of figure 2.

In the PPA, Congress declared that pollution should be prevented or reduced at the source whenever feasible; pollution that cannot be prevented should be recycled in an environmentally safe manner, whenever feasible; pollution that cannot be prevented or recycled should be treated in an environmentally safe manner whenever feasible; and disposal or other release into the environment should be employed only as a last resort and should be conducted in an environmentally safe manner. Consequently, EPA expanded TRI by requiring facilities to report additional information about their efforts to reduce pollution at its source, including the quantities of TRI chemicals they manage in waste, both on- and off-site, including amounts recycled, burned for energy recovery, or treated. EPA began capturing this information on Form R in 1991, as illustrated by "Other Waste Management" on the right side of figure 2.

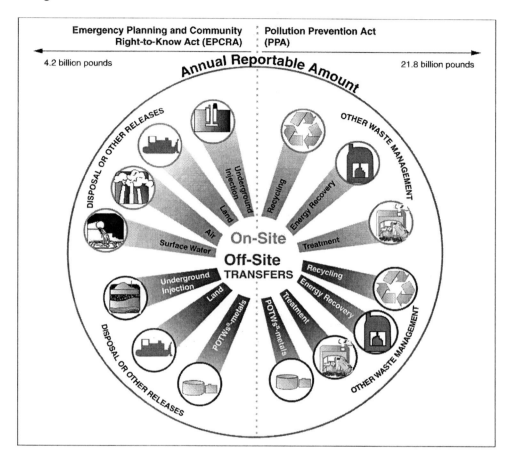

Figure 2. Types of TRI Data Reported on Form R.

Beginning in 1995, EPA allowed facilities to use a 2-page Certification Statement (Form A) to certify that they are not subject to Form R reporting for a given non-PBT chemical provided that they (1) did not release more than 500 total pounds and (2) did not manufacture, process, or otherwise use more than one-million total pounds of the chemical. Form A contains the facility identification information found on Form R and basic information about the identity of the chemical being reported. However, Form A does not contain any of the Form R details about quantities of chemicals released or otherwise managed as waste.

Beginning with Reporting Year 2001, EPA has provided the Toxics Release Inventory–Made Easy software (TRI-ME) to assist facilities with their TRI reporting. TRI-ME leads prospective reporters interactively through a series of questions that eliminate a good portion of the analysis required to determine whether a facility needs to comply with the TRI reporting requirements, including the threshold calculations needed to determine Form A eligibility. If TRI-ME determines that a facility is required to report, the software provides guidance for each of the data elements on the reporting forms. The software also provides detailed guidance for each step through an integrated assistance library. Prior to submission,

TRI-ME performs a series of validation checks before the facility prints the forms for mailing, transfers the data to diskette, or submits the information electronically over the Internet.

Each year, EPA compiles the TRI reports and stores them in a database known as the Toxics Release Inventory (TRI). In 2004—the latest year for which data are publicly available—23,675 facilities filed a total of nearly 90,000 reports, including nearly 11,000 Form As. In total, facilities reported releasing 4.24 billion pounds of chemicals to the environment and handling 21.8 billion pounds of chemicals through other waste management activities.

EPA recently embarked on a three-phase effort to streamline TRI reporting requirements and reduce the reporting burden on industry. During the first phase, EPA removed some data elements from Form A and Form R that could be obtained from other EPA information collection databases to simplify reporting. As part of the second phase, EPA issued the TRI Burden Reduction Proposed Rule, which would have allowed a reporting facility to use Form A for (a) non-PBT chemicals, so long as its releases or other disposal were not greater than 5,000 pounds, and (b) for PBT chemicals when there are no releases or other disposal and no more than 500 pounds of other waste management (e.g., recycling or treatment). The phase III changes that EPA was considering proposing would have allowed alternate-year reporting, rather than yearly reporting. The phase II and III changes generated considerable public concern that they will negatively impact federal and state governments' and the public's access to important public health information.

EPA Does Not Appear to Have Followed Internal Guidelines in All Respects When Developing TRI Rule

Although we have not yet completed our review, our preliminary observations are that EPA does not appear to have followed its own rulemaking guidelines in all respects when developing the new TRI reporting requirements. Throughout the rule development process, senior EPA management generally has the discretion to depart from the guidelines, including by accelerating the development of the proposed regulations. Nevertheless, we discovered several significant differences between the guidelines and the process EPA followed in this case: (1) late in the rulemaking process, senior EPA management directed consideration of a burden reduction option that the TRI workgroup had considered but which had subsequently been dropped from consideration; (2) EPA developed this option on an expedited schedule that appears to have provided a limited amount of time for conducting various impact analyses; and (3) the expedited schedule afforded little, if any, time for internal stakeholders to provide input to senior EPA management about the impacts of the proposal during Final Agency Review.

First, the TRI workgroup charged with identifying options to reduce reporting burdens on industry identified three possible options for senior management to consider. The first two options allowed facilities to use Form A in lieu of Form R for PBT chemicals, provided the facility has no releases to the environment. Specifically, the workgroup considered and analyzed options to facilities to:

- report PBT chemicals using Form A if they have zero releases and zero total other waste management activities; or
- report PBT chemicals using Form A if they have zero releases and no more than 500 pounds of other waste management activities.

The third option was to create a form, in lieu of Form R, for facilities to report "no significant change" if their releases changed little from the previous year.

According a June 2005 briefing for the Administrator and interviews with senior EPA officials, the Office of Management and Budget (OMB) had suggested increasing the Form A eligibility for non-PBT chemicals from 500 to 5,000 pounds as a possible burden reduction option. However, the TRI workgroup had previously dropped that option from consideration. In fact, EPA's economic analysis—dated July 2005—did not evaluate the impact of raising the Form A reporting threshold because the TRI workgroup pursued the "no significant change" option. Nonetheless, by the time the TRI burden reduction proposed rule was published in October 2005, it included the option to increase Form A reporting eligibility from 500 to 5,000 pounds.

Second, although we could not determine from the documents EPA provided or the discussions we held with EPA officials what actions the agency took between the June 2005 briefing for the Administrator and the October 2005 publication of the TRI proposal in the *Federal Register*, the Administrator provided direction after the briefing to expedite the process in order to meet a commitment to OMB to provide burden reduction by the end of December 2006. Subsequently, EPA staff worked to revise the economic analysis to consider the impact of raising the Form A reporting threshold. However, that analysis was not completed before EPA sent the proposed rule to OMB for review and was only completed just prior to the proposal being signed by the Administrator on September 21, 2005 and ultimately published in the *Federal Register* for public comment on October 4, 2005.

Third, it appears that EPA management received limited input from internal stakeholders, including the TRI workgroup, after directing that the proposed rule include the option to increase the Form A reporting threshold from 500 to 5,000 pounds. EPA conducted a Final Agency Review burden reduction proposal, as provided for in the internal rulemaking guidelines. Final Agency Review is the step where EPA's internal and regional offices would have discussed with senior management whether they concurred, concurred with comment, or did not concur with the final proposal. It appears that the review pertained to the "no significant change" option rather than increased threshold option. As a result, the EPA Administrator or EPA Assistant Administrator for Environmental Information likely received limited input from internal stakeholders about the increased Form A threshold prior to sending the TRI Burden Reduction Proposed Rule to OMB for review and publication in the *Federal Register* for public comment.

Finally, in response to the public comments to the proposed rule, nearly all of which were negative, EPA considered alternative options and revised the rule to allow facilities to report releases of up to 2,000 pounds on Form A. We continue to review EPA documents and meet with EPA officials to understand the process EPA followed in developing the TRI burden reduction proposal. We expect to have a more complete picture for our report in June.

IMPACT OF REPORTING CHANGES ON INFORMATION AVAILABLE TO THE PUBLIC IS LIKELY TO BE SIGNIFICANT

We believe that the impact of EPA's changes to the TRI reporting requirements will likely have a significant impact on environmental information available to the public. While our analysis confirms EPA's estimate that the TRI reporting changes could result less than 1 percent of total pounds of chemical releases no longer being included in the TRI database, the impact on information available to some communities is likely to be more significant than these national aggregate totals indicate. EPA estimated that these reports amount to 5.7 million pounds of releases not being reported to the TRI (only 0.14% of all TRI release pounds) and an additional 10.5 million pounds of waste management activities (0.06% of total waste management pounds). Examined locally, the impact on data available to some communities is likely to be more significant than these national totals indicate. To understand the potential impact of EPA's changes to TRI reporting requirements at the local level, we used 2005 TRI data to estimate the number of detailed Form R reports that would no longer have to be submitted in each state and the impact this could have on data about specific chemicals and facilities. We provide a summary of our methodology and estimates of these impacts, by state, in Appendix I. In addition, preliminary results from our January 2007 survey of state TRI coordinators indicate that they believe EPA's changes to TRI reporting requirements will have, on balance, a negative impact on various aspects of TRI, including environmental information available to the public.

We estimated that a total of nearly 22,200 Form R reports could convert to Form A if all eligible facilities choose to take advantage of the opportunity to report under the new Form A thresholds. The number ranges by state from 25 Form Rs in Vermont (27.2 percent of Form Rs in the state) to 2,196 Form Rs in Texas (30.6 percent of Form Rs in the state). As figure 3 shows, Arkansas, Idaho, Nevada, North Dakota, and South Dakota could lose less than 20 percent of the detailed forms, while Alaska, California, Connecticut, Georgia, Hawaii, Illinois, Maryland, Massachusetts, New Jersey, New York, North Carolina, Rhode Island, and Texas could lose at least 30 percent of Form R reports.

For each facility that chooses to file a Form A instead of Form R, the public would no longer receive detailed information about a facility's releases and waste management practices for a specific chemical that the facility manufactured, processed, or otherwise used. While both Form R and Form A capture information about a facility's identity, such as mailing address and parent company, and information about a chemical's identity, such its generic name, only Form R captures detailed information about the chemical, such as quantity disposed or released onsite to air, water, and land or injected underground, or transferred for disposal or release off-site. Form R also provides information about the facility's efforts to reduce pollution at its source, including the quantities managed in waste, both on- and off-site, such as amounts recycled, burned for energy recovery, or treated. We provide a detailed comparison of the TRI data on Form R and Form A in Appendix II.

One way to characterize the impact of the TRI reporting changes on publicly available data is in terms of information about specific chemicals at the state level. The number of chemicals for which no information is likely to be reported under the new rule ranges from 3 chemicals in South Dakota to 60 chemicals in Georgia. That means that all quantitative information currently reported about those chemicals could no longer appear in the TRI database. Figure 4 shows that thirteen states—Delaware, Georgia, Hawaii, Iowa, Maryland, Massachusetts, Missouri, North Carolina, Oklahoma, Tennessee, Vermont, West Virginia, and Wisconsin—could no longer have quantitative information for at least 20 percent of all reported chemicals in the state.

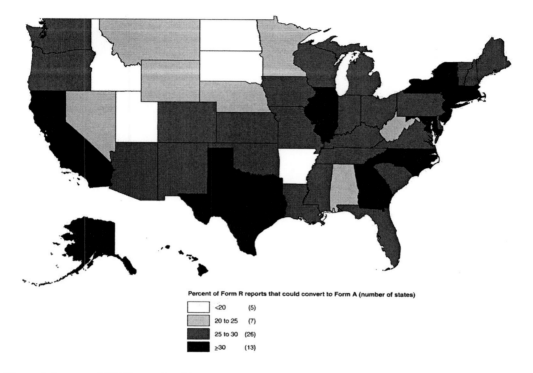

Figure 3. Impact of TRI Reporting Changes on Number of Form R Reports.

The impact of the loss of information from these Form R reports can also be understood in terms of the number of facilities that could be affected. We estimated that 6,620 facilities nationwide could chose to convert at least one Form R to a Form A, and about 54 percent of those would be eligible to convert all their Form Rs to Form A. That means that approximately 3,565 facilities would not have to report any quantitative information about their chemical releases and other waste management practices to the TRI, according to our estimates. The number of facilities ranges from 5 in Alaska to 302 in California [3]. As an example, one of these facilities is ATSC Marine Terminal—a bulk petroleum storage facility in Los Angeles County, California. In 2005, it reported releases of 13 different chemicals—including highly toxic benzene, toluene, and xylene—to the air. Although the facility's releases totaled about 5,000 pounds, it released less than 2,000 pounds of each chemical. As figure 5 shows, more than 10 percent of facilities in each state except Idaho would no longer have to report any quantitative information to the TRI. The most affected states are Colorado, Connecticut, the District of Columbia, Hawaii, Massachusetts, and Rhode Island, where more than 20 percent of facilities could choose to not disclose the details of their chemical releases and other waste management practices. Furthermore, our analysis found that citizens living in 75 counties in the United States—including 11 in Texas, 10 in Virginia, and 6 in Georgia—could have no quantitative TRI information about local toxic pollution.

Figure 4. Impact of TRI Reporting Changes on Number of Chemicals Reported on Form R.

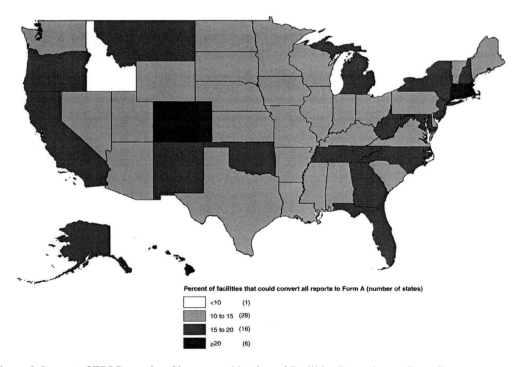

Figure 5. Impact of TRI Reporting Changes on Number of Facilities Reporting on Form R.

The Environmental Protection and Community Right-to-Know Act requires that facilities submit their annual TRI data directly to their respective state, as well as to EPA. Last month, we surveyed the TRI program contacts in the 50 states and the District of Columbia to gain their perspective on the TRI, including an understanding of how TRI is used by the states. We also asked for their beliefs about how EPA's increase in the Form A eligibility threshold would affect TRI-related aspects in their state, such as information available to the public, efforts to protect the environment, emergency planning and preparedness, and costs to facilities for TRI reporting.

Although our analysis of the survey is not final, preliminary results from 49 states and the District of Columbia show that the states generally believe that the change will have a negative impact on various aspects of TRI in their states [4]. Very few states reported that the change will have a positive impact. The states reported that the TRI changes will have a negative impact on such TRI aspects as information available to the public and efforts to protect the environment. Specifically, 23 states—including California, Maryland, New York, and Oklahoma—responded that the changes will negatively impact information available to the public, 14 states—including Louisiana, Ohio, and Wyoming—reported no impact, and one state, Virginia, reported a generally positive impact. Similarly, 22 states responded that the change will negatively impact efforts to protect the environment, 11 reported no impact, and 5 said it will have a positive impact. States also responded that raising the eligibility threshold will have no impact on TRI aspects such as emergency planning and preparedness efforts and the cost to facilities for TRI reporting. For example, 22 states responded that the change will have no impact on the cost to facilities for TRI reporting, 12 said it will have a positive impact, and no states said it will have a negative impact. The totals do not always sum to 50 because some states responded that they were uncertain of the impact on some aspects of TRI.

Finally, we evaluated EPA's estimates of the burden reduction impacts that the new TRI reporting rules would likely have on industry's reporting costs, the primary rationale for the rule changes. EPA estimated that the TRI reporting changes will result in an annual cost savings of approximately $5.9 million. (See table 1.) This amounts to about 4 percent of the $147.8 million total annual cost to industry, according to our calculations.

Table 1. EPA Estimates of Annual Savings from Changes to TRI Reporting Requirements

Option	Newly eligible Form Rs	Eligible facilities	Burden (hours per form)	Annual burden savings (hours)	Cost savings per form	Annual cost savings
New PBT chemical eligibility	2,360	1,796	15.5	36,480	$748	$1,764,969
Increased eligibility for non-PBT chemicals	9,501	5,317	9.1	86,924	438	4,160,239
Total	**11,861**	**6,670**		**123,404**		**$5,925,208**

Source: EPA based on reporting year 2004 TRI data.

This amounts to an average savings of less than $900 annually for each facility. EPA also projected that not all eligible facilities will chose to use Form A, based on the agency's experience from previous years. Furthermore, according to industry groups, much of the reporting burden comes from the calculations required to determine and substantiate Form A eligibility, rather than from the amount time required to complete the forms. As a result, EPA's estimate of nearly $6 million likely overestimates the total cost savings (i.e., burden reduction) that will be realized by reporting facilities.

We are continuing to review EPA documentation and meet with EPA officials to understand the process they followed in developing the TRI burden reduction proposal. We expect to have a more complete picture for our report later this year.

A SYSTEM TO TRACK PERCHLORATE SAMPLING AND CLEANUP RESULTS IS STILL NEEDED

Perchlorate is a salt that is easily dissolved and transported in water and has been found in groundwater, surface water, drinking water, soil, and food products such as milk and lettuce across the country. Health studies have shown that perchlorate can affect the thyroid gland and may cause developmental delays during pregnancy and early infancy. In February 2005, EPA established a new safe exposure level, or reference dose, for perchlorate, equivalent to 24.5 parts per billion in drinking water [5]. However, EPA has not established a national drinking water standard, citing the need for more research on health effects. As a result, perchlorate, like other unregulated contaminants, is not subject to TRI reporting. In May 2005 we issued a report that identified (1) the estimated extent of perchlorate found in the United States; (2) what actions the federal government, state governments, and responsible parties have taken to clean up or eliminate the source of perchlorate; and (3) what studies of the potential health risks from perchlorate have been conducted and, where presented, the author's conclusions or findings on the health effects of perchlorate.

Perchlorate has been found by federal and state agencies in groundwater, surface water, soil, or public drinking water at almost 400 sites in the United States. However, because there is not a standardized approach for reporting perchlorate data nationwide, a greater number of sites than we identified may already exist in the United States. Perchlorate has been found in 35 states, the District of Columbia, and 2 commonwealths of the United States, where the highest concentrations ranged from 4 parts per billion to more than 3.7 million parts per billion. (At some sites, federal and state agencies detected perchlorate concentrations as low as 1 part per billion or less, yet 4 parts per billion is the minimum reporting level of the analysis method most often used.) More than 50 percent of all sites were found in California and Texas, and sites in Arkansas, California, Texas, Nevada, and Utah had some of the highest concentration levels. However, roughly two-thirds of sites had concentration levels at or below 18 parts per billion, the upper limit of EPA's provisional cleanup guidance, and almost 70 percent of sites had perchlorate concentrations less than 24.5 parts per billion, the drinking water concentration calculated on the basis of EPA's recently established reference dose (see figure 6).

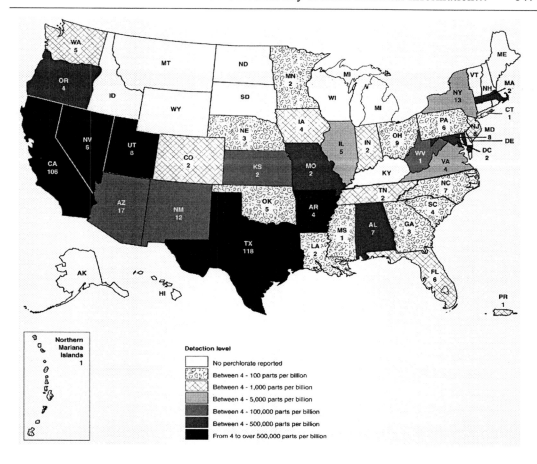

Figure 6. Maximum Perchlorate Concentrations Reported in any Media and Number of Sites, January 2005.

At more than one-quarter of the sites, propellant manufacturing, rocket motor testing, and explosives disposal were the most likely sources of perchlorate. Public drinking water systems accounted for more than one-third of the sites where perchlorate was found. EPA sampled more than 3,700 public drinking water systems and found perchlorate in 153 systems across 26 states and 2 commonwealths of the United States. Perchlorate concentration levels found at public drinking water systems ranged from 4 to 420 parts per billion. However, only 14 of the 153 public drinking water systems had concentration levels above 24.5 parts per billion. EPA and state officials told us they had not cleaned up these public drinking water systems, principally because there was no federal drinking water standard or specific federal requirement to clean up perchlorate. Further, EPA currently does not centrally track or monitor perchlorate detections or the status of cleanup activities. In fact, several EPA regional officials told us they did not always know when states had found perchlorate, at what levels, or what actions were taken. As a result, it is difficult to determine the extent of perchlorate in the United States or the status of cleanup actions, if any.

Although there is no specific federal requirement to clean up perchlorate or a specific perchlorate cleanup standard, EPA and state environmental agencies have investigated, sampled, and cleaned up unregulated contaminants, such as perchlorate, under various federal environmental laws and regulations. EPA and state agency officials have used their authorities under these laws and regulations, as well as under state laws and action levels, to sample and clean up and/or require the sampling and cleanup of perchlorate by responsible parties. For example, according to EPA and state officials, at least 9 states have established non-regulatory action levels or advisories, ranging from under 1 part per billion to 18 parts per billion. Where these action levels or advisories are in effect, responsible parties have been required to sample and clean up perchlorate. Further, certain environmental laws and programs require private companies to sample for contaminants, which can include unregulated substances such as perchlorate, and report to environmental agencies. According to EPA and state officials, private industry and public water suppliers have generally complied with regulations requiring sampling for contaminants and agency requests to sample or clean up perchlorate. DOD has sampled and cleaned up when required by specific environmental laws and regulations but has been reluctant to sample on or near active installations, unless a perchlorate release due to DOD activities is suspected and a complete human exposure pathway is likely to exist.

Finally, EPA, state agencies, and/or responsible parties are currently cleaning up or planning cleanup at 51 of the almost 400 sites where perchlorate has been found. The remaining sites are not being cleaned up for a variety of reasons. The reason most often cited by EPA and state officials was that they were waiting for a federal requirement to do so.

We identified and summarized 90 studies of perchlorate health risks published since 1998. EPA and DOD sponsored the majority of these studies, which used experimental, field study, and data analysis methodologies. For 26 of the 90 studies, the findings indicated that perchlorate had an adverse effect. Eighteen of these studies found adverse effects on fetal or child development resulting from maternal exposure to perchlorate. Although the studies we reviewed examined whether and how perchlorate affected the thyroid, most of the studies of adult populations were unable to determine whether the thyroid was adversely affected. Adverse effects of perchlorate on the adult thyroid are difficult to evaluate because they may happen over longer time periods than can be observed in a research study. However, adverse effects of perchlorate on fetal or child development can be studied and measured within study time frames. We also found some studies considered the same perchlorate dose amount but identified different effects. The precise cause of the differences remains unresolved but may be attributed to an individual study's design type or the physical condition of the subjects, such as their age. Such unresolved questions are one of the bases for the differing conclusions among EPA, DOD, and academic studies on perchlorate dose amounts and effects.

In January 2005, NAS issued its report on the potential health effects of perchlorate. The NAS report evaluated many of the same health risk studies included in our review. NAS reported that certain levels of exposure may not adversely affect healthy adults but recommended that more studies be conducted on the effects of perchlorate exposure in children and pregnant women. NAS also recommended a perchlorate reference dose, which is an estimated daily exposure level from all sources that is expected not to cause adverse effects in humans, including the most sensitive populations. The reference dose of 0.0007 milligrams per kilogram of body weight is equivalent to a drinking water exposure level of 24.5 parts per billion, if all exposure comes from drinking water. In January 2006, EPA issued guidance stating that this exposure level is a preliminary cleanup goal for environmental cleanups involving perchlorate.

We concluded that EPA needed more reliable information on the extent of sites contaminated with perchlorate and the status of cleanup efforts, and recommended that EPA work with the Department of Defense, other federal agencies and the states to establish a formal structure for better tracking perchlorate information. In December 2006, EPA reiterated its disagreement with the recommendation stating that perchlorate information already exists from a variety of other sources. However, we found that the states and federal agencies do not always report perchlorate detections to EPA and as a result EPA and the states do not have the most current and complete accounting of perchlorate as an emerging contaminant of concern. We continue to believe that the inconsistency and omissions in the available data that we found during the course of our study underscore the need for a more structured and formal system, and that such a system would serve to better inform the public and others about the locations of perchlorate releases and the status of clean ups.

PRELIMINARY OBSERVATIONS

Contrary to EPA's assertions, in our view EPA's recent changes to the Toxics Release Inventory significantly reduce the amount of information available to the public about toxic chemicals in their communities. EPA's portrayal of the potential impacts of the TRI reporting rule changes in terms of a national amount of pollution is quite misleading and runs contrary to the legislative intent of EPCRA and the principles of the public's right-to-know. TRI is designed to provide states and public citizens with information about the releases of toxic chemicals by facilities in their local communities. Citizens drink water from local sources, spend much of their time on land near their homes and places of business, and breathe the air over their local communities. We believe that the likely reduction in publicly availability data about specific chemicals and facilities in local communities should be considered in light of the relatively small cost savings to industry afforded by the TRI reporting changes.

Madam Chairman, this concludes my prepared statement. I would be happy to respond to any questions that you and Members of the Committee may have.

APPENDIX I:
GAO ESTIMATES OF THE IMPACT OF REPORTING CHANGES ON TRI DATA

We analyzed 2005 TRI data provided by EPA to estimate the number of Form Rs that could no longer be reported in each state and determine the possible impacts that this could have on data about specific chemicals and facilities [6]. Table 2 provides our estimates of the total number of Form Rs eligible to convert to Form A, including the percent of total Form Rs submitted by facilities in each state. The table also provides our estimates of the number of unique chemicals for which no quantitative information would have to be reported in each state, including the percent of total chemicals reported in each state. The last two columns provide our estimates for the number of facilities that would no longer have to provide quantitative information about their chemical releases and waste management practices, including the percent of total facilities reporting in each state.

Table 2. Estimated Impact of TRI Reporting Changes on Number of Form Rs, Chemicals, and Facilities, by State

State	Form Rs		Chemicals		Facilities	
	Number	Percent of total	Number	Percent of total	Number	Percent of total
AK	59	36.6	8	17.0	5	15.6
AL	456	22.0	34	17.1	69	12.9
AR	247	17.7	18	5.8	39	11.0
AZ	221	27.7	12	10.8	50	15.0
CA	1,533	37.5	36	18.2	302	19.9
CO	162	25.8	11	11.1	51	21.8
CT	299	33.5	16	15.4	73	20.6
DC	4	28.6	2	18.2	2	28.6
DE	80	27.7	24	23.3	10	14.1
FL	479	27.4	19	13.2	119	17.2
GA	678	30.9	60	29.1	132	16.7
HI	67	37.9	12	26.1	9	23.1
IA	371	27.7	34	22.2	46	10.6
ID	41	14.4	8	10.4	8	7.3
IL	1,155	30.0	37	16.4	171	14.3
IN	900	25.6	29	14.6	143	14.4
KS	291	28.3	23	16.0	41	14.0
KY	490	25.7	28	15.3	63	13.4
State	**Number**	**Percent of total**	**Number**	**Percent of total**	**Number**	**Percent of total**
LA	665	25.6	34	13.1	46	12.4
MA	574	38.0	23	20.4	119	20.1
MD	221	32.6	24	22.6	34	16.6
ME	105	26.1	8	11.3	14	13.7
MI	965	29.7	36	19.0	145	16.1
MN	263	21.0	20	15.4	55	11.5
MO	498	27.3	43	21.7	80	14.2
MS	265	25.0	29	18.7	37	11.8
MT	61	21.8	10	13.5	7	15.2
NC	705	30.1	43	24.9	148	17.8
ND	29	13.8	7	11.5	6	12.5
NE	116	20.3	11	7.9	24	12.9
NH	98	29.1	13	17.3	23	16.1
NJ	582	35.1	34	16.0	101	19.3
NM	96	29.2	11	15.3	15	19.2
NV	96	21.2	14	18.9	19	14.3
NY	663	31.8	33	19.1	122	17.2
OH	1,557	28.5	38	12.6	218	13.8
OK	273	26.1	30	23.3	50	15.2
OR	236	28.6	16	15.5	47	15.5
PA	1,253	29.9	30	15.2	192	14.9
RI	112	39.3	12	17.4	30	23.4
SC	596	29.0	36	17.6	78	15.0
SD	44	19.6	3	5.8	10	10.5
TN	569	27.6	40	20.9	105	16.2
TX	2196	30.6	29	9.3	210	14.1
UT	146	19.9	11	9.9	25	12.6
VA	401	25.2	23	14.8	70	14.3
VT	25	27.2	9	23.7	6	14.6
WA	276	26.4	22	19.8	43	12.5
WI	692	25.4	31	21.2	113	12.5
WV	222	22.8	40	24.1	35	17.4
WY	60	23.6	9	14.5	5	10.9
Total	**22,193**				**3,565**	

APPENDIX II:
COMPARISON OF TRI DATA ON FORM R AND FORM A

Form R	Form A
Facility Identification Information	Facility Identification Information
• TRI Facility ID Number	• TRI Facility ID Number
• Reporting year	• Reporting year
• Trade secret information (if claiming that toxic chemical is trade secret)	• Trade secret information (if claiming that toxic chemical is trade secret)
• Certification by facility owner/operator or senior management official	• Certification by facility owner/operator or senior management official
• Facility name, mailing address	• Facility name, mailing address
• Whether form is for entire facility, part of facility, federal facility, or contractor at federal facility	• Whether form is for entire facility, part of facility, federal facility, or contractor at federal facility
• Technical contact name, telephone number, Email address	• Technical contact name, telephone number, Email address
• Public contact name, telephone number	
• Standard Industrial Classification (SIC) code	• Standard Industrial Classification (SIC) code
• Dun and Bradstreet number	• Dun and Bradstreet number
• Parent company information (name, Dun and Bradstreet number)	• Parent company information (name, Dun and Bradstreet number)
Chemical Specific Information	Chemical Specific Information
• Chemical Abstracts Service (CAS) registry number	• Chemical Abstracts Service (CAS) registry number
• EPCRA Section 313 chemical or chemical category name	• EPCRA Section 313 chemical or chemical category name
• Generic name	• Generic name
• Distribution of each member of the dioxin or dioxin-like compound category	
• Generic name provided by supplier if chemical is component of a mixture	
• Activities and uses of the chemical at facility, whether chemical is: • produced or imported for on-site use/processing, for sale/distribution, as a byproduct, or as an impurity • processed as a reactant, a formation component, article component, repackaging, or as an impurity • otherwise used as a chemical processing aid, manufacturing aid, or as an ancillary or other use	
• Maximum amount onsite at any time during the year	

Form R	Form A
On-site Chemical Release Data	**On-site Chemical Release Data**

•   Quantities released on-site to:
    • air as fugitive or non-point emissions
    • air as stack or point emissions
    • surface water as discharges to receiving streams or water bodies (including names of streams or water bodies)
    • underground injection
    • land, including RCRA Subtitle C landfills, other landfills, land treatment/application farming, RCRA Subtitle C surface impoundments, other surface impoundments, other land disposal
 •   Basis for estimates of releases (i.e., monitoring data or measurements, mass balance calculations, emissions factors, other approaches)
 •   Quantity released as a result of remedial actions, catastrophic events, or one-time events not associated with production processes
	Not reported on Form A

On-site Chemical Waste Management Data	**On-site Chemical Waste Management Data**

•   Quantities managed on-site through:
    • recycling
    • energy recovery
    • treatment
 •   Recycling processes (e.g., metal recovery by smelting, solvent recovery by distillation)
 •   Energy recovery methods (e.g., kiln, furnace, boiler)
 •   Waste treatment methods (e.g., scrubber, electrostatic precipitator) for each waste stream (e.g., gaseous, aqueous, liquid non-aqueous, solids)
 •   On-site waste treatment efficiency
	Not reported on Form A

Off-site Transfers for Release or Other Waste Management	**Off-site Transfers for Release or Other Waste Management**

•   Quantities transferred to any Publicly Owned Treatment Works(POTW)
    • POTW name(s), address(es)
 •   Quantities transferred to other location for disposal or other release
    • underground injection
    • other land release
 •   Quantities transferred to other location for waste management
    • treatment
    • recycling
    • energy recovery
 •   Quantity transferred off-site for release, treatment, recycling, or energy recovery that resulted from remedial actions, catastrophic events, or one-time events not associated with production processes
	Not reported on Form A

Form R	Form A	Form R
•	Off-site location(s) name and address	
•	Basis for estimates for amounts transferred	
•	Whether receiving location(s) is/are under control of reporting facility/parent company	

	Source Reduction and Recycling Activities	Source Reduction and Recycling Activities
•	Total quantities, for (1) the prior and (2) current reporting years and estimated totals for (3) the following and (4) second following years for: • on-site disposal to underground injection wells, RCRA Subtitle C landfills, and other landfills • other on-site disposal or other releases • off-site transfer to underground injection wells, RCRA Subtitle C landfills, and other landfills • other off-site disposal or other releases • on-site treatment • on-site recycling • on-site energy recovery • off-site treatment • off-site recycling • off-site energy recovery	Not reported on Form A
•	Production ratio or activity index	
•	Source reduction activities the facility engaged in during the reporting year (e.g., inventory control, spill/leak prevention, product modifications)	
•	Option to submit additional information on source reduction, recycling, or pollution control activities	

Sources: EPA TRI Form R and Form A.

REFERENCES

[1] http://www.epa.gov/triexplorer and http://www.epa.gov/enviro

[2] These facilities were not included in the original manufacturing industries, but EPA began requiring TRI reports from seven new industries—including electric utilities that burn coal and/or oil for the purpose of generating electricity—starting in 1998.

[3] Appendix I provides the number of affected facilities for each state.

[4] Survey results from those states responding as of February 1, 2007.

[5] The reference dose of 0.0007 milligrams per kilogram of body weight per day is equivalent to 2 liters of drinking water per day containing 24.5 parts per billion of perchlorate when consumed by an adult weighing 70 kilograms (or 154 pounds), assuming that all perchlorate exposure comes from drinking water.

[6] The EPA anticipates issuing the 2005 TRI Public Data Release in April, 2007.

In: Environmental Research Advances
Editor: Peter A. Clarkson, pp. 155-170

ISBN: 978-1-60021-762-3
© 2007 Nova Science Publishers, Inc.

Chapter 7

BIOMASS PRODUCTION WITH FAST-GROWING TREES ON AGRICULTURAL LAND IN COOL-TEMPERATE REGIONS: POSSIBILITIES, LIMITATIONS, CHALLENGES

*Martin Weih** and Nils-Erik Nordh*

Swedish University of Agricultural Sciences,
Department of Crop Production Ecology,
P.O. Box 7043, SE-750 07, Uppsala, Sweden

ABSTRACT

High density plantations of fast-growing tree species grown on fertile land are today a viable alternative for the production of bio-fuels in many countries of cool-temperate regions. For example in Sweden, willow biomass plantations are commercially grown for energy purpose. Rapid development of sustainable production systems and infrastructure, along with recent progress in tree crop breeding resulted in high biomass production potential on fertile agricultural land. In addition, tree plantations can increase biodiversity in open agricultural landscapes and serve as tools for the amelioration of environmental problems at local (e.g., waste product contamination through phytoremediation) and global scale (e.g., increased greenhouse effect through carbon sequestration). Thus, multifunctional biomass plantations offer additional possibilities in terms of, e.g., wastewater cleaning and carbon sequestration. However, the establishment and large-scale implementation of woody biomass plantations is often controversial due to, for example, presumed negative influences on biodiversity and the cultural heritage landscape, along with negative public attitudes. Conflicting interests of different parts of society and socio-political issues (e.g., agricultural and energy policy, market developments, public attitudes) are therefore major barriers for the rapid development of woody bio-fuel plantations in Sweden and many other countries, rather than climatic, technical or environmental constraints. In future, careful analysis of the non-technical, non-climatic barriers at regional level and possibly the development of guidelines for the

---

* Corresponding author (M. Weih): phone +46-67 25 43, FAX +46-67 28 90, e-mail Martin.Weih@vpe.slu.se

establishment and sustainable, environmentally friendly management of woody biomass plantations could be means to boost the utilization of woody biomass from agricultural land in many countries of the world.

INTRODUCTION

High density plantations of fast-growing tree species, grown in single stem or coppice systems and producing woody biomass, are today a viable alternative to food production on agricultural land (c.f. reviews by Mitchell et al. 1992; Zsuffa et al. 1993; Makeschin 1999). The biomass produced is used for pulp and paper industries as well as energy purposes, but also as construction wood or fodder. Compared with traditional forest trees such as spruce or pine, the species used (e.g., eucalypts, poplars, aspens and willows) have a number of biological characteristics in common which make them amendable to intensive culture, i.e., high productivity rates particularly during the first years after planting, high photosynthetic capacity in combination with large leaf areas and high allocation of biomass to harvestable stem wood in relation to root and leaf biomass (e.g., Cannell et al. 1988; Ceulemans et al. 1996). The establishment of plantations of fast-growing trees has more in common with agricultural crops than forestry. Ground preparation is carried out using conventional agricultural machinery and methods. In general, the crop is planted by simply pushing stem cuttings into the cultivated soil. Shoots and roots quickly develop from these cuttings. The developing shoots are left to grow on for between 2 and 4 years (coppice systems) or 10 to 20 years (single tree systems) before being harvested with specialised machinery (Figure 1). In the case of coppice systems, the stump of each stool is left in the ground and produces more shoots that grow for a further 2 to 4 years until the next harvest. Several 2- to 4-year cutting cycles can take place in coppice systems before yield declines and the crop needs to be replaced.

Especially when grown for energy purposes, intensively managed plantations of fast-growing trees on agricultural land are gaining increasing interest in many countries of the world, due to their efficient and sustainable land use in combination with an increasing demand for renewable energy sources (Volk et al. 2004) and the additional possibilities for environmental control (phytoremediation, Crites 1984; Aronsson and Perttu 2001). For example, life cycle analysis of willow biomass production on agricultural land based on commercial clones and cultivation practices in Sweden revealed high energy efficiency of the production system: The produced biomass yields 20 times more energy compared to the energy input, a figure which is more than 3 times higher than the energy balance (i.e., energy output / energy input) of annual energy crops such as wheat and rape (Börjesson 2006). In addition, plantations of fast-growing trees require a reduced chemical input when compared to conventional arable crops, can provide an alternative use of agricultural land and can enhance the local environment through increased biodiversity. Thus, agricultural land cropped with fast-growing trees can become an important supplier of energy with marginal contribution to the increasing quantity of greenhouse gases in the atmosphere. These plantations can sequester large amounts of carbon (for *Salix* see Grogan and Matthews 2002) and may have a great potential to contribute to carbon-managed future economies in terms of fossil fuel replacement and carbon sequestration (Hall and House 1994; Tuskan and Walsh 2001).

Figure 1. Combined harvest and chipping of 3-year old shoots in a 7-year old Salix stand near Enköping, Central Sweden. Harvest is performed during winter when the ground is frozen and the plants are dormant. Photo: P. Aronsson.

The greatest potential for high biomass yields with fast-growing trees is commonly assumed in areas with warm and humid climatic conditions supporting rapid plant growth, e.g., the warm-temperate and sub-tropical climate. The most common tree species used under these climate conditions are eucalypts and poplars, for which adequate production systems have been developed (e.g., in central and southern Europe, North America, Australasia, Brasilia). But also cool-temperate and boreal floras hold fast-growing tree species that are suitable for intensively managed fast-wood plantations in high-latitude regions. Thus, adequate production systems were developed also for boreal regions (Weih 2004). In Sweden, for example, a production system for the commercial production of biomass from intensively managed willow has been developed in the frame of a National Energy Forestry Programme and currently willow plantations for energy purpose cover an area of more than 16 000 ha in Sweden. However, intensive willow culture has raised concerns about environmental, biodiversity and landscape design issues, and there are risks for conflicts between the interests of commercial willow growers and, e.g., environmentalists. In this chapter, an overview of the development of a biomass production system based on willow will be followed by a discussion of environmental and landscape design issues. Further, an analysis of potential synergies and conflicts between the objectives of commercial willow culture, nature conservation and landscape design will be presented.

DEVELOPMENT OF COMMERCIAL BIOMASS PRODUCTION BASED ON WILLOWS IN SWEDEN

Willows (genus *Salix*) are part of the natural flora in temperate-arctic regions and products from willows have been used traditionally in Sweden. The cultivation of willows (basket making, fencing, etc.) was performed at a rather small scale until relatively recently, when the potential of willow as bioenergy crop came into focus. Today Sweden is one of the few countries in the world where willow cultivation exceeds poplar growing.

In Sweden, research on willows for the purpose of biomass production was firstly initiated in the late 1960s in response to a predicted shortage of raw material for the pulp and paper industry. The energy crisis in the early 1970s supplied further arguments for the development of national energy systems that are not based on fossil fuels. In addition, the existence of many district heating systems in Sweden favored the large-scale utilization of woody biomass (Hoffmann and Weih 2005). Thus, in Sweden, willow cultivation as a source of biomass for energy purpose has been developed within the frame of a National Energy Forestry Programme (Sirén et al. 1987, Christersson et al. 1993) and expanded from a few hectares around 1970 to about 16 000 ha by the end of the 1990s. An increase to about 200 000 ha, which corresponds to almost 10 % of the agricultural land in Sweden, is envisaged during the next 15 years (Energimyndigheten 2001).

The research and development of willow culture in Sweden involved many research groups dealing with all aspects of willow genetics, physiology and ecology. The majority of research on the willow production system was done at the Swedish University of Agricultural Sciences (SLU). For example, the typical design of the willow production system in densely-spaced double-rows was developed at the SLU (Sirén et al. 1987, Christersson et al. 1993) and the research resulted in the publication of management guidelines for practical willow culture on agricultural land (e.g., Danfors et al. 1998). Also nature conservation aspects were addressed along with the development of the willow production system (Gustafsson 1987, Aronsson 1995). The plant material used for the first willow trials often performed poorly in terms of growth rate and frost tolerance. A country-wide campaign in 1978 resulted in 190 *Salix* varieties with annual shoots longer than 2.5 m (Sennerby-Forsse et al. 1983). New collections have been made since then and the clone archive at the SLU today holds about 600 different varieties of *Salix*. Based on these collections and early breeding work by G. Sirén, breeding programs for *Salix* with the aim of increasing biomass production by phenotypic screening and recurrent selection have been initiated at the SLU (Gullberg, 1993). As a result, genetic linkage maps are now available for *Salix* (Rönnberg-Wästljung 2001, Tsarouhas et al. 2002, Rönnberg-Wästljung et al. 2003).

The research and development of willow culture in Sweden was accompanied by commercialization of both production system and breeding. Thus, most commercial willow plantations in Sweden are today administrated by the Federation of the Swedish Farmers Coops, through the company Lantmännen Agroenergi AB located in central Sweden (Larsson et al. 1998). This enterprise has contacts with subcontractors and utility operators and guarantees the proper handling of the crop through advisory service. The company takes also care of harvest and delivery of the willow wood chips to the closest district heating plant.

Species of *Salix*, as well as closely related *Populus*, are characterized by high initial growth rate, but relatively short life time; easy vegetative propagation (stem cuttings; e.g., willows) and/or formation of coppice sprouts or suckers (e.g., aspens); relatively early flowering; small genome size; and light wood that is uniform in texture and used by industry and for energy (e.g., Dickmann 2001, Verwijst 2001). Easy propagation, early flowering and small genome size are all characteristics that facilitate rapid improvements by means of breeding efforts and make willows and poplars very interesting for plant breeders (Taylor 2002, Wullschleger et al. 2002).

A major characteristic of production systems based on fast-growing trees is a very rapid achievement of maximum production figures already after few years (e.g., Weih 2004): biomass production rates (mean annual increment, MAI) of > 20 Mg ha^{-1} year^{-1} were achieved after 3 to 5 years in fertilized (experimental) willow stands grown in Sweden and Canada (Christersson 1986, Labrecque and Teodorescu 2003). However, such high production figures have been achieved only in small-scale experimental plantations and biomass yields in commercial willow plantations grown in southern and central Sweden are usually in the range 5 to 10 Mg ha^{-1} year^{-1}.

The fertilization with mineral nutrients is normally necessary to ensure sufficiently high biomass yields for the economically sound culture of fast-growing trees such as willows, especially when grown under cool-temperate climate (Weih 2004). In Sweden, the use of nutrient-rich residues as an alternative, cost-efficient fertilization method was developed in parallel with the development of appropriate production systems for willow culture as a source of biomass for energy purpose (Perttu 1993, Aronsson and Perttu 2001). The principal residues used in Sweden are urban wastewater, landfill leachate, log-yard runoff, sewage sludge and wood-ash and the benefits are both environmental and economic (Dimitriou and Aronsson 2005).

ENVIRONMENTAL CONCERNS, BIODIVERSITY AND LANDSCAPE DESIGN

During recent years the planting of large areas of fast-growing trees has sparked off much controversy, especially in the areas of the world where most of the expansion of fast-wood plantations is expected to take place (i.e., South America and East Asia). Critics of these plantations of fast-growing trees include environmentalists, who argue that they are causing harm to wildlife, water resources and the soil and decrease biodiversity (Cossalter and Pye-Smith 2003). Indeed, much tree planting in recent decades, especially in the Tropics, has had negative environmental impacts, but the problems of plantation forests are frequently associated with inappropriate location and management; therefore, relatively modest changes in the ways that tree plantations are established and managed could result in greatly improved outcomes for the environment (WWF 2003). Thus, the localization and management of plantations of fast-growing trees should be planned carefully, in order to reduce the risks for environment and utilise at best their potential to improve environmental and landscape quality. For example, intensively managed plantations of fast-growing trees imply the risk of negative effects on the environment in terms of herbicide application, deteriorated soil properties and nutrient leakage. However, when compared to conventional arable crops, the perennial character of the crops used in tree plantations commonly leads to reduced herbicide application (i.e., herbicide treatment is required only during the establishment of the crop).

Further, afforestation with fast-growing trees on former arable land has often positive effects on soil properties due to less frequent use of heavy machinery, especially when the use of harvesters is restricted to the period of frozen ground in winter-cold regions. For example, carbon sequestration and water holding capacity were found to increase in formerly arable soils that were planted with fast-growing willow and poplar for 6 to 10 years (Kahle et al. 2005). The perennial nature of tree plantations also leads to less nutrient leakage compared to conventional arable crops provided that the rate and timing of fertilization is adjusted to the nutrient delivery and nutrient holding capacity of the soil in relation to the uptake capacity of the trees (Ericsson et al. 1992). This may imply the application of repeated, small fertilizations throughout the growing season according to a scenario following tree growth, i.e., small amounts early and late in the season and larger amounts in mid-summer (Ingestad 1987, Alriksson et al. 1997). Particularly in plantations with wider tree spacing grown in single-tree systems (e.g., poplars and eucalypts), nutrient supply by placing close to the roots of individual trees is an option to prevent leakage to groundwater (van den Driessche 1999). If tree plantations are managed according to these principles, they are unlikely to endanger the environment by means of groundwater contamination even under the short periods of plant nutrient uptake that are typical under the cool climate of high-latitude regions (Aronsson et al. 2000).

In contrast to the potential risks of nutrient leakage from fertilized plantations, plantations of fast-growing trees can offer great possibilities for environmental control at a local scale in terms of, e.g., phytoremediation. Thus, based on the large nutrient quantities taken up by fast-growing trees, the plantations can be used as recipients for municipal wastewater and industrial sludge and simultaneous biomass production (multifunctional biomass plantations, Perttu 1993, Isebrands and Karnosky 2001, Aronsson and Perttu 2001, Labrecque and Teodorescu 2003). A major problem occurs particularly in regions with extended dormant periods during winter, where only very small quantities of nutrients are taken up by plants. This sort of wastewater treatment can therefore be accomplished in winter-cold regions only if the wastewater is stored in ponds or lagoons during the winter period (Rosenqvist et al. 1997).

Tree plantations can have positive or negative effects on biodiversity, depending on location, management and previous land use (Cossalter and Pye-Smith 2003) and studies on biodiversity in plantations of fast-growing trees often arrive at contradictory conclusions especially when different kinds of organisms are considered (Hartley 2002). Thus, the landscape context (forest or agricultural, Hanowski et al. 1997, Weih et al. 2003) along with the land-use type that the plantation replaces (Christian et al. 1994) and spatial scale aspects (e.g. size or shape of plantation; Christian et al. 1994, Berg 2002) influence the impact of establishment of tree plantations on biodiversity. In addition, many animals use multiple habitats and therefore depend on certain habitat combinations (With et al. 1997, Law and Dickman 1998) and the diversity of land use types in a given landscape can have large impacts on biodiversity.

Compared to natural mixed coniferous forests of the Pacific Northwest, Halpern and Spiess (1995) regarded large poplar plantations to decrease flora diversity in the long term. In comparison to managed coniferous forests and farmland in boreal Sweden, young poplar and willow plantations have however been concluded to increase vascular-plant diversity (Figure 2; Gustafsson 1987, Weih et al. 2003, Augustson et al. 2006).

Figure 2. July aspect inside a 2-year old Salix stand grown on the island Gotland, south-eastern Sweden. Small-scale plantations of fast-growing trees can enhance biodiversity at landscape-scale. Photo: M. Weih.

Similar to the observations on floras, fauna diversity (birds and mammals) is frequently found to be higher in willow and hybrid poplar stands compared to agricultural croplands (Christian et al. 1998, Berg 2002). Thus, the more extensive management of tree plantations compared to intensively managed cereal crops can improve habitat quality for many organisms including plants, birds and arachnids (Berg 2002, Blick and Burger 2002, Dhondt and Wrege 2003, Weih et al. 2003). In addition, plantations of fast-growing trees appear to have a potential as important habitats for gamebirds (Sage and Robertson 1994). The location of tree plantations within the landscape context exerts a strong influence on their effect on biodiversity: A native wood nearby the tree plantation will facilitate the migration of birds, insects and plants and, thus, increase biodiversity. If no native woods are present, small groups of native trees can be planted nearby the fast-wood plantation. The native woods will serve as retreat area where animals and insects can survive during the periods of clear-cut in the fast-wood plantation. In large plantations, different parcels of the stand can be established and harvested in different years in order to enhance structural diversity and biodiversity. The same goal could be achieved by establishing a large number of smaller plantations, which are harvested in different years. Both alternatives can possibly improve the possibilities for biological control of herbivorous insects, which can be a serious problem in fast-wood plantations (Björkman et al. 2004). Mixtures of different species or varieties can be established either by planting them in different blocks across the field or random mixtures within each row (e.g., Aronsson 1995, Volk et al. 2004). These mixtures increase structural and functional diversity and contribute to reduce the impact of pests and diseases (Ramstedt 1999, McCracken and Dawson 2001). Finally, a few meters wide stripes along the edges of tree plantations could be left without trees to support the development of a rich native flora and fauna.

It hence appears that intensively managed tree plantations, if not to large in size and if planned creatively, can increase the abundance and diversity of many organisms, particularly in landscapes dominated by either conventional agricultural fields or managed coniferous forests, by increasing structural diversity. However, plantations should be avoided close to open habitats of high conservation values such as certain types of wet meadows and buffer zones along streams, because such sites often have a conservation value linked to openness of the habitat (Berg 2002). With the above restrictions in mind, small-scale tree plantations (or sub-divided larger plantations) grown on agricultural land may have the potential to positively affect biodiversity in many farmland regions, where the set-aside of agricultural land from food production suggests the development of alternative uses.

Plantations of fast-growing trees may already after few years of growth be up to 6 or 8 m in height and it is important to consider the impact this will have on the landscape at the early planning stages of a fast-wood plantation. Many concerns are raised regarding the impact of plantations of fast-growing trees on the landscape (e.g., Skärbäck and Becht 2005): Will the open agricultural landscape be "darkened" by a new type of forest? How will the new crop be adapted to the traditional field crops of agricultural landscapes? Will we soon experience a total transformation of our ingrained cultural heritage? Studies regarding the effects of plantations of fast-growing trees on visual landscape perception usually arrive at different conclusions, e.g., the more open the landscape is, the larger can the size of the plantations be and particularly small plantations placed in open agricultural landscape can enrich landscape perception; small-scale undulating landscape only "tolerates" small plantations (Rode 2005, Skärbäck and Becht 2005). A general comment is often that plantations of fast-growing trees are an exciting new feature of the landscape, these plantations can enhance the aesthetic value of landscape by adding variation (e.g., autumn colors; Figure 3) and structure in homogeneous agricultural landscape (Skärbäck and Becht 2005).

Figure 3. Variation of autumn colours in 1-year old shoots of different Salix varieties grown in a plantation near Uppsala, Central Sweden. Cultures of fast-growing trees can increase the aesthetic value of landscape, particularly if different varieties are planted. Photo: N-E. Nordh.

If used creatively as part of active landscape analysis and design, plantations of fast-growing trees can greatly improve the visual and recreational values of a landscape (e.g., Enberg 2002, Rode 2005). For example, monotonous structures in blocks of tree plantations can be broken up by edge stripes of native herbs and woods, which are flowering and thereby can increase the aesthetic value of the landscape (Rode 2005).

Concerns have also been raised regarding the impact that plantations of fast-growing trees might have on archeological sites and deposits. However, these plantations will be no more damaging to archaeology than arable cropping or commercial forestry and most critical sites such as very wet profiles are not desirable for the establishment of fast-growing trees even by other (biological) reasons. Particular attention should nevertheless be paid to the potential impact of plantations of fast-growing trees on the conservation of historic, designed landscape and it is important to consider these aspects at the early planning stages of this kind of plantation.

SYNERGIES AND CONFLICTS OF INTERESTS

Plantations of fast-growing trees are economic enterprises in most cases and economic profit arguments (i.e., maximised biomass yield after shortest possible time) usually guide their localization, shape and management in the first place. Localization, shape and management of commercial fast-wood plantations are partly conflicting with the primary interests of nature conservation and landscape design (Table 1). Regarding localization of fast-wood plantations, no serious conflicts of interest appear to exist between the interests of commercial growers, nature conservationists and landscape designers (Table 2) if plantations are established outside areas of high nature conservation and cultural heritage values. Synergies between commercial and nature conservational interests could be achieved by localization of plantations nearby native woods, which favour biodiversity (see above) and also shelter the plantation from extremes in climate (e.g., frost) and, thereby, improve sustainable productivity of the plantation (Weih 2006). No serious conflicts of interests are expected between nature conservation and landscape design principles of shaping and managing tree plantations (Tables 1, 2). Possibly could synergies be achieved through the establishment of edge habitats consisting of attractive (flowering) plants that have high conservational value for plants and insects.

Commercial willow culture may frequently conflict with the interests of nature conservation and landscape design regarding the shape and management of tree plantations. For example, large monoclonal, block-shaped and very fast growing tree stands that are harvested frequently as a whole are desirable from a purely commercial perspective, whereas many small-sized, unregularly shaped stands consisting of several species and/or clones and harvested in different years would be preferred by nature conservationists and landscape designers (Tables 1, 2). An active dialogue between commercial *Salix* growers, nature conservationists and landscape designers will facilitate feasible compromises between the conflicting interests.

Misconceptions and a lack of knowledge on intensive tree culture and its effects on environment and landscape are among the most common non-technical barriers for the full-scale implementation and dissemination of plantations of fast-growing trees in Sweden and other countries (IEA 2005, Weih 2006).

Lack of correct information, ignorance of best-practice advice (Helby et al. 2006) and misconceptions often generate negative public attitudes and biased decisions of authorities that are directly or indirectly involved in the formal regulation process regarding these plantations (Weih 2006).

Table 1. Desirable actions for localization, shape and management of plantations of fast-growing trees according to purely commercial, environmental (nature conservation) and landscape design principles (after Weih 2006)

	Commercial	Nature conservation	Landscape design
Localization	Agricultural land, short transport distances to end user, easy accessibility with heavy machinery	Agricultural land, nearby native woods, not in areas of high nature conservation value and not in landscape dominated by forest	Not in areas of high cultural heritage value and at visually exposed locations where traditional landscape views would be destroyed, preferably in open and homogeneous agricultural landscape, not in areas dominated by forest (to prevent a negative perception of landscape, the total fraction of forested area in a landscape should not exceed 40 - 50%, Wöbse 2003)
Shape	Large and block-shaped plantations that can be harvested effectively, homogeneous and highly productive plant material (preferably monoclonal stands) that rapidly generates homogeneous biomass	Small-scale plantations, stripes or small groups of native woods in the edges of the plantation, heterogeneous plant material (preferably multi-species or polyclonal stands) including male-sex varieties (insect pollination!)	Small-scale plantations, stripes of native vegetation consisting of attractive species (flowering, autumn colors, etc.)
Management	Effective weeding (chemical) prior to plantation establishment, optimal mineral nutrient fertilization "according to plant demand" (preferably with sludge or waste water from sewage plant), short cutting cycles, simultaneous harvest of the whole stand	Restrictive application of chemical weeding, not to high nutrient fertilization "according to plant demand" (to prevent leakage to groundwater), long cutting cycles, different parts of plantation are harvested different years, native edge habitats are left untouched at least during harvest of the tree plantation	No deep ploughing (to prevent disturbance of archeological deposits), native edge habitats are left untouched at least during harvest of the tree plantation

Table 2. Summary of possible conflicts of interest (yes, no) between commercial (C), environmental (nature conservation, N) and landscape design (L) principles with regard to the localization, shape and management of plantations of fast-growing trees grown on agricultural land (after Weih 2006)

	Localization		Shape		Management	
	N	L	N	L	N	L
C	no	no	yes	yes	yes	yes
N		no		no		no

There is indeed a large body of knowledge available since many years on how fast-wood plantations should be localized and managed in order to favor environmental and landscape qualities (e.g., Aronsson 1995, DEFRA 2002, Volk et al. 2004). However, this knowledge is rarely applied in commercial plantations of fast-growing trees. Campaigns and guidelines to disseminate valid information on fast-wood culture and its effects on the environment to authorities and the public, as well as improved communication among authorities, will help to remedy misconceptions and lack of correct information.

The above section has shown that there are potential negative impacts from commercial fast-wood plantations on environment and landscape, but these may be controlled and minimised by application of a creative and integrated attitude to ensure that establishment and management of fast-wood plantations operate in an optimal way to secure the positive impacts on rural economic activity, nature conservation (biodiversity) and cultural heritage.

CONCLUSION

Plantations of fast-growing trees *(Salix, Populus)* offer great possibilities for the efficient use of agricultural land in cool-temperate regions. If the biomass produced is used as bio-fuel, the plantations have a great potential to contribute to carbon managed future economies, because they contribute only marginally to the production of atmospheric greenhouse gases. By combining biomass production and phytoremediation in tree plantations, waste products from society (waste water, sludge, ash) can be used as resources to improve tree growth and generate added values in terms of both environment and economy. Plantations of fast-growing trees grown on agricultural land can improve biodiversity at landscape level, in particular if the plantations are established instead of cultures of cereals and spruce or fallow ground in homogeneous agricultural landscape. These tree plantations can also positively affect soil properties compared to conventional agriculture. Particularly plantations of relatively small size offer great possibilities for landscape design, because they are an exciting new feature in most regions and can enhance the aesthetic value of landscape by adding variation and structure.

Limitations for the localization, shape and management of fast-growing tree plantations are set by economic constraints, the landscape context and by environmental concerns. Regarding the localization of plantations, no serious conflicts between commercial and environmental interests are expected if the plantations are planned outside areas of high nature or culture conservation value. Synergy effects are possible if plantations are located nearby native woods. With respect to plantation management, the commercial practice of intensive tree culture might frequently conflict with the interests of nature and culture conservationists. An active dialogue between commercial growers and nature/culture conservationists will facilitate feasible compromises between the two sides.

Major barriers for the full-scale implementation and dissemination of fast-wood plantations in cool-temperate regions are non-technical, such as lack of correct information and misconceptions in the public and at regulating authorities. A major challenge for the future is to remove these barriers, along with improvement of the plant material used in the plantations.

REFERENCES

Alriksson, B., Ledin, S. and Seger, P. (1997). Effect of nitrogen fertilization on growth in a *Salix viminalis* stand using a response surface experimental design. *Scandinavian Journal of Forest Research*, 12, 321-327.

Aronsson, P. (1995). *Energiskogsodling och naturvårdshänsyn*. Uppsala, Sweden: Swedish University of Agricultural Sciences. 15 p. (in Swedish)

Aronsson, P. G., Bergström, L. F. and Elowson, S. N. E. (2000). Long-term influence of intensively cultured short-rotation willow coppice on nitrogen concentrations in groundwater. *Journal of Environmental Management*, 58, 135-145.

Aronsson, P. and Perttu, K. (2001). Willow vegetation filters for wastewater treatment and soil remediation combined with biomass production. *Forestry Chronicle*, 77, 293-299.

Augustson, Å., Lind, A. and Weih, M. (2006). Floristic diversity in willow biomass plantations. *Svensk Botanisk Tidskrift*, 100, 52-58. (in Swedish, with English Summary)

Berg, Å. (2002). Breeding birds in short-rotation coppices on farmland in central Sweden - the importance of *Salix* height and adjacent habitats. *Agricultural Ecosystems and Environment*, 90, 265-276.

Björkman, C., Bommarco, R., Eklund, K. and Höglund, S. (2004). Harvesting disrupts biological control of herbivores in a short-rotation coppice system. *Ecological Applications,* 14, 1624-1633.

Blick, T. and Burger, F. (2002). Arachnids (Arachnida: Araneae, Opiliones, Pseudoscorpiones) at a short-rotation coppice experimental plot. *Naturschutz und Landschaftsplanung*, 34, 276-284. (In German, with English summary)

Börjesson, P. (2006). *Life cycle assessment of willow production*. Lund, Sweden: Lund University, Dept. of Technology and Society, Report no. 60. 21 p. (In Swedish, with English Abstract)

Cannell, M. G. R., Sheppard, L. J. and Milne, R. (1988). Light use efficiency and woody biomass production of poplar and willow. *Forestry*, 61, 125-136.

Ceulemans, R., McDonald, A. J. S. and Pereira, J. S. (1996). A comparison among eucalypt, poplar and willow characteristics with particular reference to a coppice, growth-modelling approach. *Biomass and Bioenergy*, 11, 215-231.

Christersson, L. (1986). High technology biomass production by Salix clones on a sandy soil in southern Sweden. *Tree Physiology*, 2, 261-277.

Christersson, L., Sennerby-Forsse, L. and Zsuffa, L. (1993). The role and significance of woody biomass plantations in Swedish agriculture. *Forestry Chronicle*, 69, 687-693.

Christian, D. P., Niemi, G. J., Hanowski, J. M. and Collins, P. (1994). Perspectives on biomass energy tree plantations and changes in habitat for biological organisms. *Biomass and Bioenergy*, 6, 31-39.

Christian, D. P., Hoffman, W., Hanowski, J. M., Niemi, G. J. and Beyea, J. (1998). *Bird and mammal diversity on woody biomass plantations in North America. Biomass and Bioenergy*, 14, 395-402.

Cossalter, C. and Pye-Smith, C. (2003). *Fast-wood forestry: myths and realities*. Jakarta, Indonesia: Center for International Forestry Research (CIFOR). 50 p. (www.cifor.cgiar.org)

Crites, R. W. (1984). Land use of wastewater and sludge. Environmental Science and Technology, 18, 140-147.

Danfors, B., Ledin, S. and Rosenqvist, H. (1998). *Short-Rotation Willow Coppice. Growers' Manual*. Uppsala, Sweden: Swedish Institute of Agricultural Engineering. 40 p.

DEFRA [Department for Environment, Food and Rural Affairs] (2002). Growing Short Rotation Coppice. *Best Practice Guidelines*. DEFRA Publication No. 7135. London: DEFRA. 32 p. (www.defra.gov.uk)

Dhondt, A. A. and Wrege, P. H. (2003). Avian biodiversity studies in short-rotation woody crops. *Final report prepared for the US Dept of Energy under cooperative agreement* No. DE-FC36-96GO10132. Ithaca, NY: Cornell University Laboratory of Ornithology.

Dickmann, D. I. (2001). An overview of the genus *Populus*. In: D. I. Dickmann, J. G. Isebrands, J. E. Eckenwalder and J. Richardson (Eds.), *Poplar Culture in North America*. Part A, Chapter 1 (pp. 1-42). Ottawa, Canada: NRC Research Press, National Research Council of Canada.

Dimitriou, I. and Aronsson, P. (2005). Willows for energy and phytoremediation in Sweden. *Unasylva*, 221 (56), 47-50.

Enberg, A. K. (2002). Övergång och förening: ett projekt för energiskog och människa i Mälardalen. Alnarp, Sweden: Swedish University of Agricultural Sciences, *Examensarbeten inom landskapsarkitektprogrammet* 2002: 17. 59 p. (in Swedish)

Energimyndigheten (2001). Energimyndighetens Klimatrapport 2001. Eskilstuna, Sweden: Swedish National Energy Administration.

Ericsson, T., Rytter, L. and Linder, S. (1992). Nutritional dynamics and requirements of short rotation forests. In: C. P. Mitchell, J. B. Ford-Robertson, T. Hinckley and L. Sennerby-Forsse (Eds.), *Ecophysiology of Short Rotation Forest Crops* (pp. 35-65). London: Elsevier.

Grogan, P. and Matthews, R. (2002). A modelling analysis of the potential for soil carbon sequestration under short rotation coppice willow bioenergy plantations. *Soil Use and Management*, 18, 175-183.

Gullberg, U. (1993). Towards making willows pilot species for coppicing production. *Forestry Chronicle*, 69, 721-726.

Gustafsson, L. (1987). Plant conservation aspects of energy forestry - a new type of land use in Sweden. *Forest Ecology and Management*, 21, 141-161.

Hall, D. O. and House, J. I. (1994). Trees and biomass energy: Carbon storage and/or fossil fuel substitution? *Biomass and Bioenergy*, 6, 11-30.

Halpern, C. B. and Spiess, T. A. (1995). Plant species diversity in natural and managed forests of the Pacific Northwest. *Ecological Applications*, 5, 913-934.

Hanowski, J. M., Niemi, G. J. and Christian, D. C. (1997). Influence of within-plantation heterogeneity and surrounding landscape composition on avian communities in hybrid poplar plantations. *Conservation Biology*, 11, 936-944.

Hartley, M. J. (2002). Rationale and methods for conserving biodiversity in plantation forests. *Forest Ecology and Management*, 155, 81-95.

Helby, P., Rosenqvist, H. and Roos, A. (2006). Retreat from *Salix* – Swedish experience with energy crops since the 1990s. *Biomass and Bioenergy*, 30, 422-427.

Hoffmann, D. and Weih, M. (2005). Limitations and improvement of the potential utilisation of woody biomass for energy derived from short rotation woody crops in Sweden and Germany. *Biomass and Bioenergy*, 28, 267-279.

IEA [International Energy Agency] (2005). Full-scale implementation of SRC-systems: Assessment of technical and non-technical barriers. *Report for IEA Bioenergy Task 30*. 29 p. (www.shortrotationcrops.com/taskreports.htm)

Ingestad, T. (1987). New concepts on soil fertility and plant nutrition as illustrated by research on forest trees and stands. *Geoderma*, 40, 237-252.

Isebrands, J. G. and Karnosky, D. F. (2001). Environmental benefits of poplar culture. In: D. I. Dickmann, J. G. Isebrands, J. E. Eckenwalder and J. Richardson (Eds.), *Poplar Culture in North America*. Part A, Chapter 6 (pp. 207-218). Ottawa, Canada: NRC Research Press, National Research Council of Canada.

Kahle, P., Baum, C. and Boelcke, B. (2005). Effect of afforestation on soil properties and mycorrhizal formation. *Pedosphere*, 15, 754-760.

Labrecque, M. and Teodorescu, T. I. (2003). High biomass yield achieved by *Salix* clones in SRIC following two 3-year coppice rotations on abandoned farmland in southern Quebec, Canada. *Biomass and Bioenergy*, 25, 135-146.

Larsson, S., Melin, G. and Rosenqvist, H. (1998). Commercial harvest of willow wood chips in Sweden. In: *Proc. 10th European Conf. and Technol. Exhibition "Biomass for Energy and Industry"*, Würzburg, Germany, 8-11 June 1998, pp. 200-203.

Law, B. S. and Dickman, C. R. (1998). The use of mosaic habitats by vertebrate fauna: implications for conservation and management. *Biodiversity Conservation*, 7, 323-333.

Makeschin, F. (1999). Short rotation forestry in Central and Northern Europe - introduction and conclusions. *Forest Ecology and Management*, 121, 1-7.

McCracken, A. R. and Dawson, W. M. (2001). Disease effects in mixed varietal plantations of willow. *Aspects of Applied Biology*, 65, 255-262.

Mitchell, C. P., Ford-Robertson, J. B., Hinckley, T. and Sennerby-Forsse, L. (Eds.) (1992). *Ecophysiology of Short Rotation Forest Crops*. London: Elsevier. 308 p.

Perttu, K. L. (1993). Biomass production and nutrient removal from municipal wastes using willow vegetation filters. *Journal of Sustainable Forestry*, 1, 57-70.

Ramstedt, M. (1999). Rust disease on willows - virulence variation and resistance breeding strategies. *Forest Ecology and Management*, 121, 101-111.

Rode, M. (2005). Energetische Nutzung von Biomasse und der Naturschutz. *Natur und Landschaft*, 80, 403-412. (in German, with English Summary)

Rönnberg-Wästljung, A. C. (2001). Genetic structure of growth and phenological traits in *Salix viminalis*. *Canadian Journal of Forest Research*, 31, 276-282.

Rönnberg-Wästljung, A. C, Tsarouhas, V., Semerikov, V. and Lagercranz, U. (2003). A genetic linkage map of a tetraploid *Salix viminalis x S. dasyclados* hybrid based on AFLP markers. *Forest Genetics*, 10, 185-194.

Rosenqvist, H., Aronsson, P., Hasselgren, K. and Perttu, K. (1997). Economics of using municipal wastewater irrigation of willow coppice crops. *Biomass and Bioenergy*, 12, 1-8.

Sage, R. B. and Robertson, P. A. (1994). Wildlife and game potential of short rotation coppice in the UK. *Biomass and Bioenergy*, 6, 41-48.

Sennerby-Forsse, L., Sirén, G. and Lestander, T. (1983). Results from the first preliminary test with short rotation willow clones. Uppsala: Swedish University of Agricultural Sciences, *SEF Project Technical Report* 30, pp. 1-37.

Sirén, G., Sennerby-Forsse, L. and Ledin, S. (1987). Energy plantations - Short rotation forestry in Sweden. In: D. O. Hall and R. P. Overend, *Biomass* (pp. 119-143). London: John Wiley.

Skärbäck, E. and Becht, P. (2005). Landscape perspective on energy forests. – *Biomass and Bioenergy*, 28, 151-159.

Taylor, G. (2002). *Populus*: Arabidopsis for forestry. Do we need a model tree? *Annals of Botany*, 90, 681-689.

Tsarouhas, V., Gullberg, U. and Lagercranz, U. (2002). An AFLP and RFLP linkage map and quantitative trait locus (QTL) analysis of growth traits in *Salix*. *Theoretical and Applied Genetics*, 105, 277-288.

Tuskan, G. A. and Walsh, M. E. (2001). Short-rotation woody crop systems, atmospheric carbon dioxide and carbon management: A U.S. case study. *Forestry Chronicle*, 77, 259-264.

Van den Driessche, R. (1999). First-year growth response of four *Populus trichocarpa x Populus deltoides* clones to fertilizer placement and level. *Canadian Journal of Forestry Research*, 29, 554-562.

Volk, T. A., Verwijst, T., Tharakan, P. J., Abrahamson, L. P. and White, E. H. (2004). Growing fuel: a sustainability assessment of willow biomass crops. *Frontiers in Ecology and Environment*, 2, 411-418.

Verwijst, T. (2001). Willows: An underestimated resource for environment and society. *Forestry Chronicle*, 77, 281-285.

Weih, M. (2004). Intensive short rotation forestry in boreal climates: present and future perspectives. *Canadian Journal of Forestry Research*, 34, 1369-1378.

Weih, M. (2006). *Willow short rotation coppice grown on agricultural land – possibilities for improvement of biodiversity and landscape design*. Report to the Swedish Environmental Protection Agency (Dnr 802-114-04). Uppsala, Sweden: Swedish University of Agricultural Sciences (SLU). 36 p. (in Swedish, with English Summary) (http://pub-epsilon.slu.se/75/)

Weih, M., Karacic, A., Munkert, H., Verwijst, T. and Diekmann, M. (2003). Influence of young poplar stands on floristic diversity in agricultural landscapes (Sweden). *Basic and Applied Ecology*, 4, 149-156.

With, K. A., Gardner, L. H. and Turner, M. G. (1997). Landscape connectivity and population distribution in heterogeneous environments. *Oikos*, 78, 151-169.

Wullschleger, S. D., Jansson, S. and Taylor, G. (2002). Genomics and forest biology: *Populus* emerges as the perennial favorite. *Plant and Cell*, 14, 2651-2655.

Wöbse, H. H. (2003). Landschaftsästhetik. Stuttgart, Germany: *Eugen Ulmer*. 304 p. (in German)

WWF [World Wildlife Fund] (2003). WWF vision for planted forests. Report for the UNFF Intersessional Experts Meeting on the Role of Planted Forests in Sustainable *Forest Management*, 24-30 March 2003, Wellington, New Zealand. (www.maf.govt.nz/mafnet/unff-planted-forestry-meeting/conference-papers/www-vision-for-planted-forests.htm)

Zsuffa, L., Sennerby-Forsse, L., Weisgerber, H. and Hall, R. B. (1993). Strategies for clonal forestry with poplars, aspens and willows. In: M. R. Ahuja and W. J. Libby (Eds.), *Clonal Forestry II, Conservation and Application* (pp. 91-119). Berlin: Springer.

In: Environmental Research Advances
Editor: Peter A. Clarkson, pp. 171-184

ISBN: 978-1-60021-762-3
© 2007 Nova Science Publishers, Inc.

Chapter 8

ATMOSPHERIC POLLUTION

Teresa Fortoul

Insituto Nacional de Cardiologia Ignacio Chavez,
Mexico City, Mexico

SOURCES OF ATMOSPHERIC POLLUTION

The modern life has increased the sources from which particles and gases could be liberated into the atmosphere. The increase in fuel combustion as a consequence of the augmentation in vehicular transportation, wood burning, land erosion, wheel and asphalt erosion, and industrial burning products are some of the origins of these aerosols and gases. Also, smoking should be mentioned as an important source of atmospheric pollution.

Atmospheric aerosols are complex mixtures of particles directly emitted into the atmosphere and particles that are formed during gas-to-particles conversion process (Kourtrakis and Sioutas, 1996).

Gases

Some of the most important gases studied in the atmosphere are: Carbon monoxide (CO), nitrogen and sulfur oxides and hydrocarbons. The contact of some of these pollutants with sunlight produces a series of chemical reactions which results in other pollutants. The best example of these photochemical products is Ozone, which induces hyperrectivity, and decreases the ability to remove infections agents from the respiratory tract.

Combustion of fuels produces the release of CO, CO_2, sulfur dioxide, formaldehyde, hydrocarbons, nitrogen oxides and particles. All these gases have diverse health effects which affects mainly those subjects with previous respiratory or cardiovascular problems (Moeller, 1992).

Tobacco Smoke

Much has been written about the damage that tobacco smoking produce and the increase in lung cancer rates in younger smokers and women (Belania, 2007), is something that we should be worried about (Hammond, 2005). But also we should take care of the collateral damage to those who decided not to smoke, but in someway are forced to inhale second hand smoke.

Nonsmokers inhale environmental tobacco smoke (ETS) which is the combination of the sidestream smoke that is released from the cigarette's burning end with the mainstream smoke exhaled by the active smoker. The exposures of the involuntary and the active smoking differ quantitatively and qualitatively. Sidestream smoke has higher concentrations of some toxics and carcinogenic substances than the mainstream (Samet, 1991).

Tobacco smoke is a complex mixture of gases and particles which incorporate thousands of chemicals. Epidemiological studies have linked involuntary smoking with several health injuries, which mainly affects children (Table 1.), but still parents keep smoking near them. Furthermore, adults might be affected with similar affections as those referred in primary smokers (Samet, 1991).

Table 1. Health effects of involuntary smoking

• **Children**	• **General Population**
• Increased lower respiratory infections	• Increased respiratory symtoms
	• Reduced lung function
• Increased respiratory symtoms	• Asthma exacerbation
• Reducen lung growth	• Increased risk of asthma
• Increased cancer risk	• Increased risk of cardiovascular disease
• Irritation of the eyes, nose, throat and lower respiratory tract	• Increased risk of nonrespiratory cancers
• Increased risk for sudden infant death	• Earlier age of menopause
• Reduced birth weight	

Modified from Samet et al, 1991.

Ozone

This is a secondary air pollutant resulted from the interaction of ultraviolet radiation with oxides of nitrogen and with volatile organic compounds (VOCs). It is a highly reactive gas and its increase in the atmosphere has been associated with an increase in respiratory diseases. Its interaction with the epithelial lining fluid initiates a chain reaction producing reactive oxygen species (ROS), which will damage cell membranes from the respiratory epithelium. This will result in the release of inflammatory factors that will activate alveolar macrophages, increasing also blood vessels permeability inducing epithelial cells damage and edema. Neutrophils will liberate elastase, peroxidase and more ROS which will perpetuate the damage. Also, an increase in mucous secretion will be evidenced (Figure 1).

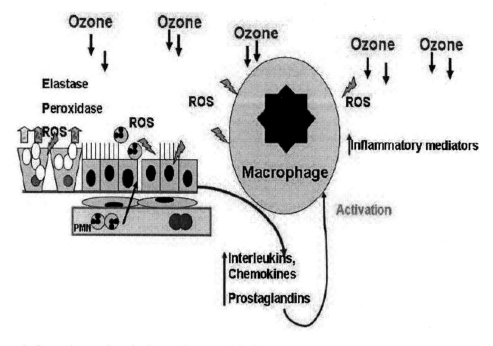

Figure 1. Ozone interactions in the respiratory epithelium.

The raise in this air pollutant has been associated with an increase susceptibility to viral and bacterial infections, also emphysema and fibrosis are associated with the exposure, and its main target area is the bronchiole.

Because ozone is a potent oxidant with substantial DNA-damage capabilities, there has been a long time concern about its possible carcinogen or co-carcinogen activity (Deblin et al, 1997).

Nitrogen Dioxide

Combustion processes generate NO and NO_2 and these gases are soluble in fluids of living organisms. When it is inhaled it is retained in the large bronchi and small airways. When it enters in contact with water in the lung it reacts to form nitric and nitrous acids. High concentrations induce edema, bronchitis, bronchiolitis and pneumonia. Affects mucociliary function, and as ozone, impaired response to infections is also reported (Samet, 1992).

Carbon Monoxide (CO)

This gas in its acute toxic action binds strongly with hemoglobin in the capillary bed resulting in a compound called carboxyhemoglobin. This gas has an affinity of two hundred times the affinity of oxygen for the available binding sites in the heme groups. The result of this strong affinity is the reduced capacity for binding oxygen and because of an allosteric change induced by its binding to the hem groups; the already fixed oxygen reduces its ability to dissociate from the heme group. It also interferes with the electron transport at mitochondrial level.

Because of the adverse effects of CO on oxygen delivery tissues, individuals with a need for high oxygen consumption or who have preexisting disorders of oxygen delivery may be highly sensitive to its effects, such as: developing fetus, the growing child, and athletes who practices strenuous exercise. Also, those with chronic hypoxemia, cardiovascular diseases and hemoglobinopathies are susceptible to the effects of CO.

CO exposure might exacerbate ischemic symptoms in patients with cardiovascular diseases as well that those with Chronic Obstructive Pulmonary Diseases (Coultas and Lambert, 1992).

Volatile Organic Compounds (VOCs)

These gases might be present in the atmosphere bound to airborne particles and are associated with smoking, auto exhaust, and several other combustion activities. Some of the compounds in this group of substances area: Benzene, tetrachloroethylen, Chloroform, Methylene chlroride, trichlorroethylene , Aromatic hydrocarbons (toluene, xylenes, ethylbencene), Alimphatic hydrocarbons (octane, decan), terpenes (limonene), Semivolatile chemicals (Chlorpyrifos, chlordane, heptachlor, diazinon, Polychlorinated biphenyls (PBCs) and Polycyclic aromatic hydrocarbons (PAHs). The main problem with these chemicals is that they share a carcinogenic or co-carcinogenic activity. One of the better known associations is benzene and leukemias (Wallace, 1991).

Suspended Particles

These pollutants have been classified according with its aerodynamic size and its composition, which is related with its source (Spengler and Wilson, 1996).

Particles identified as PM_{10}, means that its aerodynamic diameter is ≤ 10 μm (Figure 2). These are considered as coarse particles and are formed by:

- Resuspended dust, soil dusts, street dust. Coal and oil fly ash.
- Metal oxides of Si, Al, Mg, Ti, Fe, $CaCO_3$, NaCl, sea salt, pollen,
- Mold spores, and plant parts

These particles have a lifetime in the air from minutes to hours and might travel from one to 10 km.

Fine particles -0.1-1.0 μm- are integrated by:

- Sulfates Nitrates, Ammonium, Hydrogen ions, elemental carbon,
- Organic compounds, metals (Pb, Cd, V, Ni, Cu, Zn).
- Particles bound water and biogenic organics

Figures 2 A and 2B. Examples of coarse particles. A pollen grain (A) and street dusts (B) are part of PM_{10} constituents observed by scanning electron microscopy. These particles are the result of industrial dusts, construction, roads and street debris.

Its sources are combustion of coal, oil, gasoline, diesel and wood. Products of the transformation of NO_x, SO_2, biogenic organics, and high temperature processes such as the products of smelters, steel mills. Their lifetimes in the atmosphere are from days to weeks and have a large range of traveling –from hundreds to thousands kilometers (Spengler and Wilson, 1996; Morawska, L., and Zhang, J., 2002).

Coarse particles are retained in the upper part of the respiratory tract while the small ones penetrate and travel deeper in the bronchial tree as it is resumed in Table 2.

Table 2. Respiratory tract structures reached by suspended particles regarding its aerodynamic size

Particles range (µm)	Structure(s)
11-7	Nasal passages
6.9- 4.7	Pharynx
4.6- 3.3	Trachea and primary bronchi
3.2- 2.1	Secondary bronchi
2.0 - 1.1	Terminal bronchi
1.0- 0.65	Bronchioli
0.64 -0.43	Alveoli

Modified from Spengler and Wilson, 1996.

Those particles with an aerodynamic diameter of <2.5 μm ($PM_{2.5}$), called fine-particles are associated with increased all-cause and cause-specific mortality and with increased respiratory and cardiovascular morbidity. While mechanisms through which inhaled particulate matter injures the pulmonary system have been documented, understanding of the biologic processes by which particulate matter may affect the cardiovascular system remains incomplete. Epidemiologic research has identified adverse physiologic effects associated with increased particulate matter exposure in persons with cardiovascular disease, including reduced heart rate variability, increased blood pressure, cardiac arrhythmias, increased oxidative stress and inflammation, progression of atherosclerosis, as well as increase risk of strokes (Hong et al., 2002).

HEALTH EFFECTS AND AIR POLLUTANTS

In the last decades the interaction of urban air pollution and cardiovascular effects has been reiteratively evaluated. The pioneers in reveal this interaction were Dockery et al in 1993 with the study of six cities in which clearly stated the impact of Total Suspended Particles (TSP) in morbidity and mortality.

As we mentioned previously the particle size which induced more damage are those identified as $PM_{2.5}$, however more attentions is oriented to ultrafine particles (PM> $_{2.5}$), which can penetrate deeper in the lung tissue.

The specific cause by which suspended particles exerts its effect on the cardiovascular system is still in study, but several reports suggest several hypotheses (Liaou et al., 2004):

- A direct action of particles transported from the lungs to systemic circulation.
- A reflexive response of the cardiac autonomic system to direct particulate activation of chemosensitive pulmonary response.
- Nonspecific responses of the cardiac autonomic system to noxious pulmonary stress mediated by sympathetic response.
- Cumulative responses to stimulus- evoked production
- Release of inflammatory cytokines (including some interleukins and tumor necrosis factor) from pulmonary macrophages, epithelial cells or fibroblasts.

A unique mechanism by which suspended particles exert its toxic effects has not been established, and as it has been mentioned above, the release of inflammatory cytokines has been suggested. Hetlanda et al., (2004) identified that coarse and fine particles induce the secretion of different cytokines, coarse particles induce IL-8 and ultrafine particles IL-6. Also, a higher apoptotic effect was observed by ultrafine particles on the lung epithelial cells.

Other alterations reported in subjects exposed to inhaled ultrafine particles were an increase in leukocytes and platelets counts, as well as an increase in blood fibrinogen. Also pulmonary vasoconstriction was referred in resting individuals exposed to ambient air particles and ozone. An explanation for some of the changes observed in this report was the decrease in adhesion molecules found by the authors; the vasoconstriction as a consequence of a decrease in the production of NO by the endothelium, and this pulmonary vasoconstriction would delay the transit of white blood cells (Frampton, 2006).

A previous report from Churg et al., (2003) referred that in cities in which the burden of air suspended particles is high, particles tended to locate in the bronchioli remaining in the connective tissue that surrounds this structure inducing airway remodeling and fibrosis, situation which be associated with chronic airflow obstruction.

In addition, the increase incidence of lung cancer in young subjects with any previous history of smoking, but whose main area of residence were cities with air pollution problems, is another injury that should be added to the list of air pollution health problems (Green et al., 1993).

ATMOSPHERIC POLLUTION AND DNA DAMAGE

The growing demands for resources such as food, land and energy have had an impact on environmental emissions and human exposure. The consequences of this exposure on human health are of great concern and continue to be elucidated (IARC, 1992).

Strong evidence suggests that long-term exposure to fine particles is an important health risk (Vineis and Husgafvel-Pursiainen, 2005), since exposure to ambient air particulate matter (PM) is associated with pulmonary and cardiovascular diseases and cancer (Brunekreef and Holgate, 2002).

Numerous studies, carrying different experimental conditions have established the evidence for genotoxicity of air pollution. PM may elicit their effects directly or indirectly. Direct effects are inherent to their physico-chemical properties, while indirect effects are due to formation of reactive oxygen (ROS) and nitrogen species by inflammatory processes. (Greim et al, 2001, Schins et al, 2004; Vineis and Husgafvel-Pursiainen, 2005).

PM carry a complex mixture of chemical components, among them, are metals, which their status as agents that interact directly or indirectly with DNA has been established (Gandhi and Kumar 2004). Thus, the relationship between inhaled metals by PM and cancer needs to be observed.

Metals: Direct Mechanisms of DNA Alterations

One of the main mechanisms of action of metals is their potential to intercalate between DNA base pairs or by alkylating them, forming covalent DNA adducts. (Casarett and Doull, 1996). DNA adduct levels have been studied as biomarkers of exposure to many chemicals, thus, they can be used to asses risk assessment. They are found in target and non target tissues and the level of DNA adducts has been correlated with the carcinogenic potential of some metals (i.e. platinum compounds)

Adduct may form with specific DNA bases such as N7-guanine N7-adenine and O6-guanine, this may lead to the formation of crosslinks between the metal and DNA (intra and interstrand-crosslinks) intrastrand-crosslinks are responsible for most of the citotoxicity seen with these type of compounds and they are very difficult to be repaired, this represents another DNA insult that may give rise to mutations during DNA metabolism processes. (Pfhuler and Wolf 1996)

Adducts are released from DNA either spontaneously or by repair processes. Ample evidence suggests that mutations arise either during DNA replication at the damaged site or during error-prone DNA repair (Hutchinson, 1989, IARC 1992).

Metals: Indirect Mechanisms of DNA Alterations

As mention earlier, indirect effect of PM in relation to their genotoxicity is the causation of oxidative stress, this phenomena may be due to direct generation of reactive oxygen species from the surface of particles, soluble compounds such as transition metals or organic compounds, also, activation of inflammatory cells capable of generating ROS and reactive nitrogen species (Greim et al; 2001, Schins et al; 2004, Risom et al; 2005). Generation of ROS as a DNA damaging agent has been widely studied in the last decade.

DNA is considered to be a main target for ROS generated as a consequence to air pollution exposure. In order to understand the generation of DNA damage in intact cells we should assume that particles enter the nucleus and that extracellular ROS initiate free radical chain reactions that reach in some point the genetic material. (Risom et al; 2005). It is very important to mention that there are other oxidant compounds present in air particles that contribute to the oxidative stress caused by PM.

As stated earlier, particles contain soluble transition metals, such as, iron, copper, chromium and vanadium that can generate ROS through Fenton type reactions and act as catalysts by Harber–Weiss reactions (Risom et al, 2005). We can mention iron as an example, which has been purposed that diesel exhaust particles contain surface functional groups that complex iron, then, it accumulates inducing oxidative stress. (Ghio et al; 2000, Han et al; 2001).

Resulting oxidative DNA damage may be implicated in cancer risk thus, measures of ROS in different cells or tissues may serve as biomarker of early effects. (Risom et al; 2005).

All these DNA insults are of great biological importance because they can instruct us about the molecular mechanisms underlying genetic toxicity of metals that are present in PM, moreover, they represent early alterations in DNA that in combination with other factors (age, sex smoking status, individual susceptibility) may reflect the potential of some chemical agents to alter DNA representing health hazards, therefore, the use and development of biomarkers of exposure and early effect are in urgent need because of the role they can play in prevention.

Biomarkers

Various biomarkers of exposure, effect, and susceptibility of human diseases have been identified and applied for human populations. These include the molecular dosimetry of internal dose and biologically effective dose of exposure to a wide variety of environmental risk factors; the characterization of early biological effects, altered structures and functions of target organs, also preclinical lesions; in addition to the identification of genetic and acquired susceptibility to diseases (Chen et al; 2005).

Excessive generation of ROS can oxidize various cellular biomolecules. Free radicals generate a wide variety of oxidative alterations in DNA, including strand breaks and base oxidations (Nagashima et al; 1995, Dybdahl et al; 2004). One of the resulting products of oxidative DNA damage is, 8-oxo-7,8-dihydro-2'-deoxyguanosine (8-oxodG) which probably is the most studied oxidation product due to its relative ease of measurement and pre-mutagenic potential. In DNA, 8-oxodG may be formed by oxidation of guanine or incorporated during replication or repair as oxidized nucleotides (8-oxodGTP). Experimental observations, combined with the data on frequent oxidative DNA damage in lymphocytes in people exposed to urban air pollution, point to 8-oxo-dG being one of the important promutagenic lesions, therefore it is at least for now, one of the best biomarkers of exposure to oxidative damage (Vineis and Husgafvel-Pursiainen, 2005).

Cytogenetic Biomarkers

These types of biomarkers are probably the most frequently used endpoints for human population studies. Two aspects had contributed to their success: their sensitivity for measuring exposure to genotoxic agents and their key role as predictors of cancer risk (Bonassi et al; 2005). Chromosomal aberrations (CA) have value as biological dosimeters and several studies reveal the strong relationship between CA and cancer risk. For the past 20 years, population studies found that subgroups with elevated CA frequencies have significantly increased cancer incidence or significantly increased lung cancer mortality compared to the subgroup with lower CA frequencies (Hagmar et al., 1994; Bonassi et al., 1995, Au and Salama 2005). According to Hagmar and co-workers (2004), ''there is no other biomarker for general cancer risk that is applicable to healthy subjects from the general population with such a high attributable proportion.'' Different chemical agents induce different types of CA, for example: compound present in PM such as metals may produce chromatid type aberrations, so, they can also serve to infer mechanisms of action.

Another biomarker that has gained importance in the last few years is the micronucleus (MN) assay. Assuming that "carcinogens are mutagens" this test is now included in a battery of test recommended by various regulatory health organizations (EPA 1999) to screen for potential carcinogens. MN test measures the capacity of a certain compound to be clastogenic (chromosome breakage) and aneugenic (Chromosome lagging due to dysfunction of mitotic spindle). One of the aspects implicated in its success is that it is a simple, non-expensive technique used *in-vivo* and *in-vitro*. Most studies asses the incidence of MN in bone marrow cells, lymphocytes or desquamated cells from urine. (Kim et al; 2000, Krishna et al; 2000, Rosenkranz and Cinningham 2000).

In addition to CA and MN, a third biomarker of exposure is the measurement of Sister Chromatid exchanges (SCE).

Sister chromatid exchanges (SCEs) occur as a normal feature of cell division in mammalian cells. They can represent the interchange of DNA replication products at apparently homologous loci and involve DNA breakage and reunion. They involve breakage of both DNA strands, followed by an exchange of whole DNA duplexes. This occurs during S phase and is efficiently induced by mutagens that form DNA adducts or that interfere with DNA replication. The test is usually done on human peripheral blood lymphocytes. (EPA 1996).

The formation of SCEs has been correlated with the induction of point mutations, gene amplification and cytotoxicity. Although its mechanism of action and biological relevance is poorly understood, it is a technique that stills holds today.

Some test agents that are good inducers of SCE are bifunctional alkylating agents, which may be the case for some metals present in PM, in addition, it can reveal potential genotoxins at lower doses than AC, so in combination with other genotoxic end-points it may serve as a good marker to asses potential genetic risks.

Finally, when assessing human population risks it is important to pay special attention to early effects that may help to predict human health risks, rather than trying to predict carcinogenic effects about complex mixtures based on data from single agents, so, biomonitoring studies for early effects such as cytogenetic biomarkers can predict potential carcinogenic exposures.

Other Affected Systems

Central nervous system is one of the targets of air pollutants, and some reports mentioned some data about metals and organic solvents. Changes in mood and behavior with low-level expsosure include lead, mercury, manganese, carbon sulfide, methylbromide, pentaborane, ethylene glycol monoethylether and other solvents.

The symptoms that patients may complain are detailed in Table 3. These changes might be also the result of other pathologies, for this reason a detailed evaluation of the patient should be performed to support or discard the effect of air pollution (Bolla, 1991).

Table 3. Symptoms Which May Suggest Neurotoxicity

- Inability to concentrate
- Loss of memory
- Depressed mood
- Anxiety
- Restlessness
- Loss of interest in work
- Changes in libido
- General apathy
- Confusion
- Sleep disturbances
- Irritability
- Headaches
- Weakness

Ozone, as we previous mentioned, has been identified as a highly reactive gas and its main effects have been referred in lung tissue. Gonzalez-Piña and Paz (1997) reported a study in cats in which ozone exposure caused modifications in brain's biochemistry that correlates with sleep disorders. Soulagea et al. (2004) identified also peripheral metabolism modifications which might explain modifications in heart rate; also describe inhibition of catecholamine metabolism in lungs and striatum. This last effect might affect motor tasks, and is part of the actions on lung response to the exposure.

FINAL REMARKS

The extension of damages that air pollution might produce is far beyond this report. All the systems, by different means will react to pollutants, manifesting its changes as acute inflammation that might evolve to chronicity with further progression to degenerative diseases, situation that it is repercuting right now in the world wide health.

ACKNOWLEDGMENTS

Authors thank the scanning microscopy images provided by Silvia Antuna-Bizarro, Armando Zepeda-Rodriguez, and Francisco Pasos-Najera; also to Veronica Rodriguez-Mata and Judith Reyes- for the tissue process for light microscopy. Also thank Blanca R. Fortoul for editorial work. Supported in part by PAPIIT-DGAPA, UNAM IN200606.

REFERENCES

Au W. and Salama. (2005). Use of biomarkers to elucidate genetic susceptibility to cancer. *Environmental and Molecular Mutagenesis.* 45:222-228.

Belania, Ch.P., Martsb, S, Schiller, J, Mark A. Socinskid, MA (2007). Women and lung cancer: Epidemiology, tumor biology, and emerging trends in clinical research. *Lung Cancer* 55, 15—23.

Bonassi S, Abbondandolo A, Camurri L, Dal Pra' A, De Ferrari M, Degrassi F, Forni A, Lamberti L, Lando C, Padovani P, Sbrana I, Vecchio D, Puntoni R. (1995). Are chromosome aberrations in circulating lymphocytes predictive of future cancer onset in humans? Preliminary results of an Italian cohort study. *Cancer Genet Cytogenet* 79, 133–135.

Bolla, K.I. (1991). Neuropsychological assessment for detecting adverse effects of volatile organic compounds on the central nervous system. *Environmental Health Perspectives* 95: 93.98.

Bonassi, S. Ugolini, D. Kirsch-Volders, M. Strmberg, U. Vermeulen, R. and Tucker J. (2005). Human Population Studies With Cytogenetic Biomarkers: Review of the Literature and Future Prospectives. *Environmental and Molecular Mutagenesis* 45:258 270.

Brunekreef, B. and Holgate, S.T. (2002). Air pollution and health, *Lancet* 360. 1233–1242.

Casarett and J, Doull. (1996). Toxicology. U.S.A 5th edition. Ed. McGraw-Hill. 1021pp.

Chien-Jen ChenT, Lin-I Hsu, Chih-Hao Wang, Wei-Liang Shih, Yi-Hsiang Hsu, Mei-Ping Tseng, Yu-Chun Lin, Wei-Ling Chou, Chia-Yen Chen, Cheng-Yeh Lee, Li-Hua Wang, Yu-Chin Cheng, Chi-Ling Chen, Shu-Yuan Chen, Yuan-Hung Wang, Yu-Mei Hsueh, Hung-Yi Chiou, Meei-Maan Wu. (2005). Biomarkers of exposure, effect, and susceptibility of arsenic-induced health hazards in Taiwan *Toxicology and Applied Pharmacology.* 206 198– 206.

Churg, A., Brauer,M., Avila-Casado, M.C., Fortoul, T.I., Wright, J.L. (2003).Chronic Exposure to High Levels of Particulate Air Pollution and Small Airway Remodeling. *Environ Health Perspect.* 111:714–718.

Coultas D.B, Lambert W.E. *Carbon Monoxide* (1992). in Indoor air pollution. A health perspective. Eds. Samet J.M. and Spengler J.D. Baltimore, Johns Hopkins Univ Press.

Devlin R.B., Raub J.A., Folinsbee L.J.(1997). Health effects of Ozone Science and Medicine 4: 9-17.

Dybdahl, M., Risom,L., Bornholdt,J., Autrup,H., Loft,S. and Wallin,H. (2004). Inflammatory and genotoxic effects of diesel particles in vitro and in vivo. *Mutat Res.*, 562, 119–131.

EPA. U.S. Environmental protection agency. (1999). HPV Challenge Program Guidance Documents.

EPA. U.S. Environmental protection agency. (1996). Health effects test guidelines. OPPTS 870.5900.In vitro sister chromatid exchange assay.

Frampton, M.W., Stewart, J.C., Oberdörster, G., Morrow, P.E., Chalupa,D., Pietropaoli, A.P., Frasier, L.M., Donna M. Speers, D.M., Christopher Cox, Ch., Huang, Li-Shan, Utell, M.J. (2006). Inhalation of Ultrafine Particles Alters Blood Leukocyte Expression of Adhesion Molecules in Humans. *Environ. Health Perspect* 114:51–58.

Gandhi, G., and N, Kumar. (2004). DNA Damage in Peripheral Blood Lymphocytes of Individuals Residing Near a Wastewater Drain and Using Underground Water Resources. *Env. Molec. Mut.* 43:235–242.

Ghio, J.H. Richards, J.D. Carter and M.C. Madden, (2000). Accumulation of iron in the rat lung after tracheal instillation of diesel particles, *Toxicol. Pathol.* 28, 619–6.

Gonzalez-Piña, R. and Paz, C. (1997). Brain monoamine changes in rats alter short period of ozone exposure. *Neurochemical Res.* 22:63-66.

Green, L.S., Fortoul, T.I., Ponciano, G., Robles, C., Rivero, O.(1993). Bronchogenic cancer in patients under 40 years old: The experience of a Latin American country. *Chest.* 104:1477-1481.

Greim, H. Borm, P., Schins, R., Donaldson, K., Driscoll, K., Hartwig, A., Kuempel,E., Oberdorster ,G. and Speit, G. (2001). Toxicity of fibers and particles. Report of the workshop held in Munich, Germany, 26–27 October 2000. *Inhal. Toxicol.* 13, 737–754.

Hagmar L, Brogger A, Hansteen IL, Heim S, Hogstedt B, Knudsen L, Lambert B, Linnainmaa K, Mitelman F, Nordenson I. (1994). Cancer risk in humans predicted by increased levels of chromosomal aberrations in lymphocytes: Nordic study group on the health risk of chromosome damage. *Cancer Res* 54:2919–2922.

Hagmar L, Stro"mberg U, Bonassi S, Hansteen I-L, Knudsen LE, Lindholm C, Norppa H. 2004. Impact of types of lymphocyte chromosomal aberrations on human cancer risk: results from Nordic and Italian cohorts. *Cancer Res* 64:2258–2263.

Hammond, D. (2005).Smoking behavior among young adults: beyond youth prevention. *Tob. Control.* 14; 181-185.

Han, K. Takeshita and H. Utsumi, Noninvasive detection of hydroxyl radical generation in lung by diesel exhaust particles, *Free Radic. Biol. Med.* 30 (2001), pp. 516–525.

Hetlanda, R.B., Casseeb, R.F., Refsnesa, M., Schwarzea, P.E., Lag M., Boereb, A.J.F., Dybinga, E. (2004). Release of inflammatory cytokines, cell toxity and apoptosis in epithelial lung cells after exposure to ambient air particles of different size fractions. *Toxicology in Vitro* 18: 203–212.

Hong, Yun-Chul; Lee, Jong-Tae; Kim, H., Kwon, Ho-Jang (2002). Air Pollution. A New Risk Factor in Ischemic Stroke Mortality. *Stroke*.33:2165-2169.

Hutchinson F. (1989). Use of data from bacyteris to interpret data on DNA damage processing in mammalian cells. *Mut Res.* 220, 269-278.

IARC International Agency for the Research on Cancer. (1992).Mechanisms of carcinogenesis in risk identification. Eds: H. Vainio, P. Magee, D. Mcgregor and A.J. Mcmichael. No 116. 608p.

Kim B.S. Cho, M.H. and Kim H. J. (2000). Statistical analysis of in vivo rodent micronucleus assay. *Mut. Res.* 469, 233.241.

Koutrakis P and Sioutas C. (1996). Physico-Chemical properties and measurement of ambient particles. Chap. 2. in: *Particles in our air: Concentrations and Health Effects.* Eds Wilson R and Spengler J.D. Boston, Harvard University Press.

Krishna G. and Hayashi M. (2000). In vivo rodent micronucleus assay: protocol, conduct and data interpretation. *Mut. Res.* 455, 156-166.

Liao, D., Duan, Y, Whitsel E.A., Zheng, Zhi-jie , Heiss, G,Chinchilli, V.M., Lin, Hung-Mo. (2004). Association of Higher Levels of Ambient Criteria Pollutants with Impaired Cardiac Autonomic Control: A Population-based Study. *Am J Epidemiol* 159:768–777.

Moeller, D.W. (1992).Environmental Health. Cambridge, Harvard Univesristy Press.

Morawska, L., Zhang, J. (2002). Combustion sources of particles. 1. Health relevance and source signatures. *Chemosphere* 49:1045–1058.

Nagashima, M., Kasai, H., Yokota, J., Nagamachi, Y., Ichinose, T. and Sagai, M. (1995). Formation of an oxidative DNA damage, 8-hydroxydeoxyguanosine, in mouse lung DNA after intratracheal instillation of diesel exhaust particles and effects of high dietary fat and beta-carotene on this process. *Carcinogenesis,* 16, 1441–1445.

Pfuhler, S. and Wolf H.U. (1996). Detection of DNA crosslinks agents with the alkaline comet assay. *Env. Mol. Mutag.* 27: 196-201.

Risom, L., Moller, P. and Loft, S. (2005). Oxidative stress-induced DNA damage by particulate air pollution. *Mutat Res.* 592(1-2):119-37.

Routledge, H.C., Ayres, J.G., Townend, J.N. (2003). Why cardiologists should be interested in air pollution. *Heart* 89:1383–1388.

Samet J.M. *Nitrogen Oxide* (1992). in *Indoor air pollution. A health perspective.* Eds. Samet J.M. and Spengler J.D. Baltimore, Johns Hopkins Univ Press.

Schins, R.P., Lightbody ,J.H., Borm, P.J., Shi, T., Donaldson, K. and Stone, V. (2004) . Inflammatory effects of coarse and fine particulate matter in relation to chemical and biological constituents. *Toxicol. Appl. Pharmacol.* 195, 1–11.

Soulagea, Ch., Perrina,D., Cottet-Emardb, J.M., Pequignota, J., Dalmaza, Y., Pequignota, J.M. (2004). Central and peripheral changes in catecholamine biosynthesis and turnover in rats after a short period of ozone exposure. *Neurochemistry Int* 45:979–986.

Symons, J.M., Wang, L., Guallar, E., Howell, E., Dominici, F., Schwab, Ange, A.B., Samet, J., Ondov, J., Harrison S., Geyh, A. (2006). A Case-Crossover Study of Fine Particulate Matter Air Pollution and Onset of Congestive Heart Failure Symptom Exacerbation Leading to Hospitalization. *American Journal of Epidemiology* 164:421-433.

Spengler J and Wilson R.(1996). Emissions, dispersion and concentration of particles. Chap. 3. in: *Particles in our air: Concentrations and Health Effects*. Eds Wilson R and Spengler J.D. Boston, Harvard University Press.

Vineis P. and Husgafvel-Pursiainen, K. (2005). Mortality and long-term exposure to ambient air pollution: ongoing analyses based on the American Cancer Society cohort. *J Toxicol Environ Health A*, 68 (13-14): 1093-109.

Vineis P. and Husgafvel-Pursiainen, K. (2005). Air pollution and cancer: biomarker studies in human populations. *Carcinogenesis* vol.26 no.11 pp.1846–1855.

Wallace L.A. Volatile Organic Compounds. (1992). in Indoor air pollution. A health perspective. Eds. Samet J.M. and Spengler J.D. Baltimore, Johns Hopkins Univ Press.

In: Environmental Research Advances
Editor: Peter A. Clarkson, pp. 185-196

ISBN: 978-1-60021-762-3
© 2007 Nova Science Publishers, Inc.

Chapter 9

THE EU METAL PROJECT - METALS IN THE ENVIRONMENT, TOXICITY AND ASSESSMENT OF LIMITS: CADMIUM

L. Tudoreanu[1], S. Prankel[4], L. Enache[1], B. Akhmetov[2], A. Omarova[2], B. Kovacs[3], Z. Gyori[3] and C. J. C. Phillips[4,5]

[1] Faculty of Land Reclamation, University of Agronomy and
Veterinary Medicine, Bucharest, Romania
[2] Institute of Human and Animal Physiology, Almaty, Kazakhstan
[3] Central Laboratory, University of Debrecen, Hungary
[4] Department of Clinical Veterinary Medicine, University of Cambridge,
Cambridge, United Kingdom
[5] School of Veterinary Science, University of Queensland, Australia

ABSTRACT

Critical limits for cadmium in the human food chain are considered by some to have too small margins of safety and some are regularly exceeded. There is therefore widespread concern about the exposure of some sections of the population to cadmium in the human food chain, in particular from offal, which is a major depot for cadmium in livestock products. During the last decade a large number of empirical models and only a very few mechanistic-phenomenological models for food chain Cd accumulation were developed.

Linking sector-specific models for Cd accumulation in the food chain is a challenge as a large variety of modelling techniques are usually used to derive specific sub-models. The easiest models to integrate into large food-chain models are the linear and nonlinear empirical models. We used several modeling techniques for sub-model construction, but in particular meta-regression for compiling results from many trials conducted on the absorption of Cd from soil by plants, on the uptake of Cd into sheep kidney and liver, and on the relationship between Cd intake by humans and standard measures of human toxicity.

Human cadmium intake derives mainly from food sources, although cigarette smoking is a significant source. Cadmium can be present in high concentrations in some offal and accumulates in kidney and liver tissue, with a greater rate for kidney accumulation than for liver. The margin of safety between typical cadmium intakes by humans and levels associated with toxicity is smaller than for other heavy metals. We determined that there is an exponential increase in β2-microglobulin with increases in cadmium intake above 302 μg/day, which corresponds to a Provisional Tolerable Weekly Intake of 3.02 μg/kg of body weight, when a Safety Margin of 10 is included. This compares with the current level set by FAO/WHO of 7 μg/kg of body weight, which is therefore believed to be too high. Human cadmium intake could be effectively reduced by routine removal of the ovine liver and kidney from the food chain in polluted regions. Consumption of just one sheep kidney could cause an average adult person to exceed their Provisional Maximum Tolerable Daily Intake (1μg/kg body weight).

The meta-analysis of published data results was used to derive prediction equations of cadmium accumulation in sheep which are applicable to a broad set of exposure situations allowing the critical examination of cadmium in the human food chain. The product of the cadmium concentration in the feed and the duration of exposure to that feed were significant predictors of the cadmium concentration in liver and kidney. The predominance of organic rather than inorganic forms of cadmium in the feed further increased accumulation. As a result, the prime measure to decrease the risk of cadmium from animal origin adversely affecting human health should be restricting the cumulative cadmium intake of herbivores. Since reduction of maximum cadmium levels in sheep feed or of the duration of their exposure are not economically viable measures of control, routine removal of the ovine liver and kidney from the food chain in polluted regions is recommended as the best option to reduce human cadmium intake. In abattoirs, cadmium concentrations in farm animals are best monitored from hair and blood concentrations, respectively.

The meta-analyses of soil data showed that a particular care should be taken if a soil has both a high soil cadmium concentration and a low pH because this situation will favour cadmium accumulation in plants. The use of soil models for monitoring cadmium fate revealed that Cd^{2+} and Pb^{2+} concentrations in the soil solution phase of contaminated soils may be reasonably well predicted from the total soil-metal concentration and the soil pH. The amounts of cadmium taken up by plants or lost through leaching are relatively small compared with the amount added to soils through phosphate fertilizer application. There is therefore a net accumulation of cadmium in soils, particularly in the upper 15 cm. The maximum soil Cd concentrations quoted by international guidelines vary between 1 mg/kg and 3 mg/kg and pasture soils contain on average 0.4 mg Cd. Some fertilizers companies have adopted the limit of 280 mg Cd/kg P, representing 25 mg Cd/kg fertilizer in the case of single superphosphate.

Linear modelling of field/pot plant data enabled us to draw inferences on Cd accumulation in plants. There is a linear relationship between the total Cd concentration in the vegetative plant (dependent variable) and the plant dry weight, the total Cd concentration in the soil solution and the duration of exposure (time from planting to harvest) (independent variables), as well as a linear relationship between Cd concentration in the reproductive organs /grains and Cd concentration in the vegetative plant organs. No linear relationship was found between Cd concentration in bulk soil and plant organs, or between Cd concentration in the soil solution and grains/reproductive organs. Although Cd uptake influences the total concentration of Ca, Mg, K, P, Mn, Fe, Cu in plants, especially in the early stages of their development, the total Cd concentration in the vegetative plant part is not influenced by the concentration of these metals in the respective organs. For non-toxic Cd concentration levels in the plant organs, the plant mineral status does not influence Cd uptake.

Mechanistic models have been constructed to estimate mineral root uptake, which usually assume that ions are transported to roots by mass flow and diffusion and are absorbed at rates that depend on their concentration at the root surface, following

Michaelis-Menten kinetics. The accuracy of such models is not high, partly because accurate estimates of root surface area are essential but difficult to achieve. Other important parameters are the cadmium concentration in the soil solution and root growth.

Soil temperature variation is considerable but little attention has been paid to this factor, which has a major impact on cadmium influx into the plant. Other factors, such as the presence of competing ions, sewage sludge use as a fertilizer, symbiotic fungi (arbuscular mycorrhizae), and bacteria, and temperature are believed to play an important role in cadmium uptake but are mainly unquantified. Soil splash may play an important role in the contamination of crops which have the edible parts within 30-40 cm above the soil surface. The contribution of wet and dry deposition also plays an important role in heavy metal accumulation in plants, but little attention has been paid to cadmium accumulation throughout the phylloplane.

There is currently only a limited understanding and quantification of key parameters which would allow a comprehensive mechanistic model of Cd uptake by different plant genotypes to be constructed, and also that there is a limited number of empirical observations of key endpoints for an empirical model. Further work is required to understand each species' influence on plant model parameters. Survey data of field-grown plants might help to elucidate different aspects of the relationship between soil parameters, air depositions and cadmium accumulation in plants.

In order to predict cadmium and lead concentrations in air, there are well-established published models that can be used. These models take into account the complex phenomena in the atmosphere and they can make estimations of the deposition rates for Cd and Pd from data on source type, the intensity of atmospheric turbulence, the emission rate, the wind speed, the degree of phenomenon complexity, the surface structure and vegetation properties of the reception area.

Further work on these aspects is essential to facilitate the construction of effective models to control excessive Cd accumulation in the human food chain.

OBJECTIVES OF THE PROJECT

1. Define critical limits of the heavy metals lead and cadmium in human and animal food, water, air and soils.
2. Produce a computer model for cadmium and lead in the human food chain that can be utilized in specific instances in the field, such as by food quality inspectors to determine effects of emissions on food quality.
3. Define low productivity systems with close-to-zero cadmium and lead accumulation
4. *Identify significant sources of cadmium and lead contamination and means of controlling them*

Methodology

a. *Literature* collection: Initially literature was collected into a database using electronic abstracting systems e.g. CD-ROM abstracts, and printed works.
b. *Flow of information between the partners and METAL MODEL construction:* Each partner adopted the most adequate modeling technique for their

area of research in order to derive the equations needed for the construction of the consolidate METAL MODEL.

Object oriented programming (OOP) languages such as C++, are an ideal tool for modelling complex systems, enabling the development of modular structured models. Each module may therefore be independently developed, updated and run.

Thus the final model is composed of eight submodels of cadmium and lead pathways in key sector areas, written in C++ by the Romanian partner, and which are based on the equations derived/found by each partner (Figure 1).

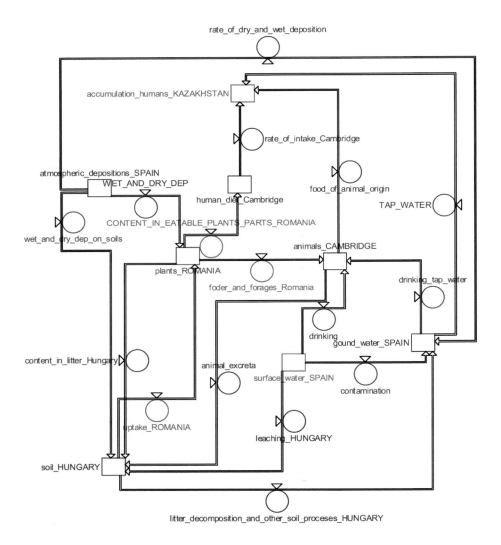

Figure 1. Flow of information between partners for the model construction. Blue – Spain and for the last two years Romania; Green – Romania; Brown –Hungary; Purple- Cambridge UK; Red –Kazakstan.

The work areas of each partner were:

1. Soils (University of Debrecen, Hungary)

The soil sub-model used a standard model (MANTEQA) for chemical equilibrium. The Hungarian team reviewed the available models for ionic transfer in soils and finally used the MANTEQA programme to model lead and cadmium behaviour in soils and to make inference on Cd and Pb concentration in the soil solution.

2. Air, Water (University of Valladolid, Spain, University of Bucharest, Romania)

The relevant data and physical parameters for Pb and Cd deposition rates and velocities were initially collected by the Spanish partner. These findings led to the identification of the most relevant equations for particle transport and deposition over different types of canopies (by the Romanian partner) and therefore enabling the link between the air models and plants models (created by the Romanian partner).

In order to create or adopt the most appropriate models for Cd and Pb in air to be linked with the plant, animal and human models, the Spanish and Romanian partner created a literature database containing information and figures for concentrations and deposition rates for Cd and Pb on the soil surface from national and international publications.

The Romanian partner identified the most relevant equations concerning the relationships between pollutants dispersion and meteorological factors and topographical factors. A simple deposition rate model was created and several published models were identified to be suitable to be included in the METAL model.

3. Plants (University of Bucharest, Romania)

The presence of cadmium in the food chain is primarily affecting humans, rather than animals or crops, because of the longevity of humans and the accumulation of cadmium through the food chain. Therefore it is essential that empirical models for plant accumulation be linked to similar models for animals and humans.

From the endogenous linear models reported, it appears that a generic approach may be possible for cadmium accumulation by crops that are subject to Michaelis-Menten kinetics, with plant cadmium concentration being positively related to cadmium in soil and negatively related to soil pH. Although first order coefficients for two species (maize and ryegrass) were different, which questions the generic approach, second order equations produced a similar negative interaction component for the two species, with the impact of soil cadmium on accumulation of cadmium therefore being dependent on soil pH and *vice versa*.

Relationships were also developed for cadmium uptake from nutrient solution. Finally, soil temperature variations of more than 15 ^{0}C occur over the growing season but little attention has been paid by researchers to this factor, which has a major impact on cadmium influx into the roots.

Although regression is known to be robust with respect to non-normality of residuals, inferences derived from the regression analyses may be questionable when non-normality of the residuals is detected.

Survey data of field-grown plants might bring into light a different aspect of the relationship between soil parameters and cadmium accumulation in plants (Smith, 1994).

Mechanistic models have been constructed to estimate mineral root uptake, which usually assume that ions are transported to roots by mass flow and diffusion and are absorbed at rates that depend on their concentration at the root surface, following Michaelis-Menten kinetics. The accuracy of such models is not high, partly because accurate estimates of root surface area are essential but difficult to achieve. Also little is known about mineral interactions in the uptake mechanisms. Other important parameters are the cadmium concentration in the soil solution and root growth. A particular disadvantage of modelling uptake in the field is the difficulty in making measurements of these parameters.

A simple endogenous deterministic model that incorporates soil and foliar uptake was proposed. Cadmium concentration in the soil solution is the main variable that links the endogenous model to soil models such as SOILCHEM or GEOCHEM. Other factors, such as the presence of competing ions and symbiotic fungi (arbuscular mycorrhizae) and bacteria and temperature plays an important role in cadmium uptake but this is largely unknown and unable to be incorporated into the endogenous model.

Sewage sludge deposition onto the leaves is known to contribute to heavy metals accumulation in plants but no attempts have been made to model sewage sludge contribution to cadmium accumulation in plants, relating this contribution to environmental factors and plant leaf physical properties.

Although soil splash may represent a source of cadmium contamination, authors investigating cadmium accumulation in plants largely omitted the contribution of soil splash, therefore modelling soil splash contribution to cadmium accumulation in plants was not considered.

The contribution of wet and dry deposition also plays an important role in heavy metal accumulation in plants, but little attention has been paid to cadmium accumulation throughout the phylloplane. Therefore the simple sub model for foliar uptake enables the estimation of maximum cadmium concentration in and onto the leaves (assuming no "loss" from the plant leaves).

The endogenous deterministic model is presented both in a modular form and also in an executable form.

The proposed deterministic model is of value for integration into the human and animal food chain model and it might be adequate for all plants using the same mechanism for cadmium uptake (Michaelis-Menten kinetics) and cadmium sequestration by phytochelatins.

Empirical models, while presenting a less satisfactory biological model compared to mechanistic models, are based on observations, which ensures a degree of accuracy that is not necessarily present in mechanistic models.

The linear regression models were finally chosen to be incorporated in the overall model and several plant models (ryegrass, soybean, maize, lettuce) were included in the soil -plant – animal-human model – the METAL model. The mechanistic models will be developed separately.

4. Animal Sources (University of Cambridge, UK)

Accumulated cadmium in the offal of ruminants consuming contaminated herbage represents a significant source of intake for some consumers (MAFF, 1997). Limits for cadmium and lead concentrations in food animal products have been established independently whereas these two toxic metals often co-exist in polluted regions. The most sensitive tissues for Cd and lead exposure were kidney, liver, hair, and teeth and regression equations were developed for the accumulation rates in these tissues. Tissue and blood cadmium concentrations increased gradually with increasing dietary lead, whereas tissue lead concentration was not sensitive to dietary cadmium, except in the ribs and heart. Using empirical data from a previous project run by the partners, we modeled the accumulation rates of the two metals in the pig body tissues and organs. This was for the most part linear over the range of dietary inclusion rates (0-2.5mgCd/Kg and 0-25mgPb/Kg) and there were also interaction between cadmium and lead, so that lead increased the accumulation in body tissues.

We used data from samples of kidney, blood, lungs, hair, heart, liver, spleen, muscle, ear, rib, skin and faeces that had been obtained from 180 fattened pigs in ten abattoirs across Hungary in a previous project. Modelling was conducted in this project to determine the optimum tissues to monitor. The samples had been analysed in triplicate for cadmium and lead concentrations by atomic absorption spectrophotometry. Highest concentrations of cadmium had been found in the kidney, followed by hair and faeces and then liver, heart and lungs. Lead concentrations were greater in the hair and faeces than other tissues. Both lead and cadmium concentrations in meat and offal were below the legal limits. There was more variation in lead than cadmium concentrations between abattoirs, and across abattoirs there was no correlation between concentrations of the two elements. Blood lead concentration was correlated with the lead concentration in bone, kidney, liver, spleen and lungs and was a better indicator of lead contamination than hair lead concentration. The cadmium concentrations of the kidney, liver, spleen, lungs and faeces were highly correlated, and it is suggested that faeces is the best on-farm indicator of cadmium contamination. Across animals, blood cadmium correlated less closely with the cadmium concentration of the body tissues than it did with the lead concentration, demonstrating positive interaction between the two elements. By contrast hair and to a lesser extent bone cadmium concentrations were negatively correlated with the lead concentration of most tissues. It is concluded that lead and cadmium concentrations in pigs are best monitored in blood and hair concentrations, respectively.

5. Human Physiochemistry and Toxicology (Institute of Human and Animal Physiology, Kazakhstan)

A dataset was derived from 60 literature sources and from 27537 peoples, (59.2%-Women, 35.1%-Men, 5.7% mixed. A total of 68.85% of the data originated in Asia, and 31.5 % from other parts of world.

Beta 2-Microglobulin (β2-Microglobulin)is a low molecular weight protein that is found in most biological fluids. It was originally isolated from urine of cadmium-poisoned patients Therefore there was a domination of data from women.

The variables recorded were CDI (Cd intake/day, DUR (Duration of exposure in years, B2MC(β2 microglobulin in urine), AGE (Age/years), CDB (Cd in blood),and CDU (Cd in urine). Independent variables were transformed as follows: CDI to 1/CDI (1/v2) and CDI**4 (v**4); DUR to 1/DUR (1/v3) and DUR**4 (v3**4); AGE to 1/AGE (1/v5) and AGE**4 (v5**4) and then we executed a fixed non-linear regression estimation. A plot of residuals against predicted values approximated to a straight line and a polynomial relationship between Beta 2 microglobulin and the independent variables CDI and DUR was found too. The same polynomial relation ship was found between blood cadmiumin urine and the independent variables CDI and AGE (R = 0.85, Standard error of estimate 1.41, P < 0.00001).

For estimating blood Pb concentration (BLC) vs daily lead intake (DLI) relationship, data from 56 trials involving more than 3469 persons living in different regions were used. Preliminary analysis showed that the age factor is important in this dose/effect relationship. Linear correlation coefficient (CC) between BLC and DLI was 0.62, while the relationship between BLC and DLI divided by age (years) gave CC 0.71. 22 reports (all of them are abstracts) have no data on mean age of the subjects being under study and 33 (abstracts too) have no standard deviation, which is important for weighting measurements in random effect modelling. WinBUGS software, which was used to construct the equations, treats these data as missing data. Running the simulation required an initial 5000 updates for the adapting phase, then 5000 for convergence reaching (burn in) and next 30000 iterations until the Monte Carlo error for each parameter of interest became less than about 5% of the sample standard deviation.

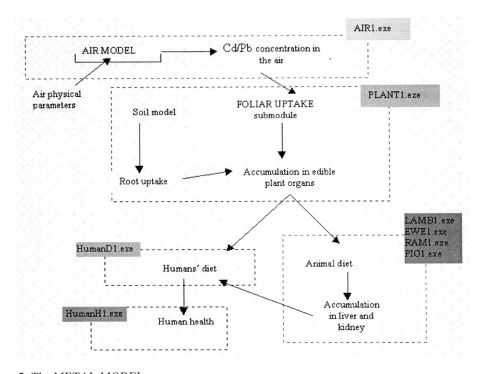

Figure 2. The METAL MODEL.

Finally the consolidated METAL MODEL was constructed, and it was composed of 8 submodels which may be run separately too. Because no data set covering the entire METAL

MODEL was available, the validation of consolidated model was not possible. However the consolidated model is constructed by linking linear models, models which do not request validation. The phenomenological plants model will be developed separately from the consolidated METAL MODEL in future collaboration between the partners

- AIR.exe - It is a simple model to calculate Cd/Pb concentration in the air. The output of this model is used for the PLANT1 model. Air models may be used to compute air Cd/Pb concentration or an appropriate database may be used too.
- PLANT.exe. - The model computes Cd uptake by plants using soil, air and plant parameters. This endogenous model can be used for monocots as well as dycots as long as the requested parameters are known. The PLANT1 model is an attempt to calculate cadmium taken up by the plant from soil solution and wet and dry aerial deposition. Cd concentration in the soil solution and cadmium concentration in air are the outputs of the soil and air model.
- LAMB1.exe .- The model computes the total Cd accumulated in lamb's liver and kidney tissues due to soil, feed and water ingestion.
- EWE1.exe - The model computes the total Cd accumulated in ewe's liver and kidney tissues due to soil, feed and water ingestion.
- RAM1.exe - The model computes the total Cd accumulated in ram's liver and kidney tissues due to soil, feed and water ingestion.
- LAMB1, EWE1, and RAM1 models are linking the plant endogenous models to the endogenous animal models.
- PIG1.exe- The model computes the total Cd concentration in pig's liver and kidney tissues due to ingested cadmium.
- HumanD1.exe - The model computes the quantity of Cd and /or Pb ingested by humans from food.
- HumanH1.exe- The model computes the Pb concentration in human blood, the Zn concentration in the human blood and the concentration of zinc protoporphyrin/hemoglobin in the human blood.

Two of the proposed four partners had collaborated in a previous Copernicus project to reduce toxic effects of heavy metals in agriculture, which brought endogenous data to utilize in the modelling process. The model was therefore constructed by exchanging information between partners in the key areas.

C) Practical research. Some practical research was undertaken to fill gaps in existing knowledge, with contributions to the soil team to support their work on speciation, which constitutes a major influence on soil transfer, to the plant team to investigate the impact of different levels of cadmium on the uptake by lettuce, and to the human team to investigate the impact of iron status on lead toxicity in children. Thus there was a focus on metal speciation because this represents a major unknown influence on metal trends are between different components in the food chain.

CONCLUSION

For humans, the toxic effects of Pb are involved in a wide spectrum of diseases. Some of the physiological processes influenced by Pb are: hem synthesis, renal functions, osteoporosis, nerve conduct velocity spermatogenesis, fetal development. Children are much more vulnerable to lead than adults due to their high rate of absorption from the digestive tract and the lungs and have a four time critical level of lead in blood (10 µg/dl) than adults, above which cognitive disturbance appears. 3500 studies over 20 years were taken into account for a meta-analisis of dose-effect relationships between food and air intake and lead accumulation in blood. Linear correlation were found having 0.93 coefficients of correlation between lead in blood and Pb uptake from food and 0.717 for lead uptake from inhaled air. The critical permissible level of lead in blood (for children) was found to be 17.8 µg/dl for dietary Pb intake and 5 µg/dl for inhaled Pb.

Cadmium is a naturally occurring heavy metal which is readily taken up by plants, and then consumed by animals and humans. It accumulates in kidney and liver, and we determined that the rate for kidney accumulation is greater than for liver. Blood cadmium as an indicator of excessive exposure was investigate by deriving nonlinear functions thorough meta-analyses, having as independent variables CDI (Cd intake/day) and AGE (Age/years). There was an exponential increase in β2-microglobulin with increases in cadmium intake above 302 µg/day, which corresponds to a Provisional Tolerable Weekly Intake of 3.02 µg/kg of body weight, when a Safety Margin of 10 is included. This compares with the current level set by FAO/WHO of 7 µg/kg of body weight, which is therefore believed to be too high. Cadmium in blood and urine were also positively related to cadmium intake and participants' age. There were two principal components of variation in the dataset, first: cadmium intake, concentrations of cadmium in blood, urine and β2-microglobulin in urine, and second: duration and age of exposure. It is concluded that the results of existing literature indicate that the current FAO/WHO Provisional Tolerable Weekly Intake for cadmium may be too high.

A meta-analysis using random effects modelling was carried out to integrate the results of 21 controlled randomized trials in which sheep were fed diets with elevated cadmium levels and cadmium concentrations in their liver and kidney were recorded after slaughter. Resulting predictions of cadmium accumulation in sheep are applicable to a broad set of exposure situations allowing the critical examination of cadmium in the human food chain. The product of the cadmium concentration in the feed and the duration of exposure to that feed were significant predictors of the cadmium concentration in liver and kidney. The predominantly organic rather than inorganic form of cadmium in the feed further increased accumulation. Other variables (dry matter intake, the vehicle of the elevated cadmium in the diet, animal age, weight and sex) were not significant. As a result, the prime measure to decrease the risk of cadmium from animal origin adversely affecting human health should be restricting the cumulative cadmium intake. Since reduction of maximum cadmium levels in sheep feed or of the duration of their exposure are not economically viable measures of control, routine removal of the ovine liver and kidney from the food chain in polluted regions is recommended as the best option to reduce human cadmium intake.

In abattoirs, Cd and Pb concentrations are best monitored from hair and blood concentrations, respectively, at least in pigs.

The meta-analyses of soil data showed that a particular care should be taken if a soil has both a high soil cadmium concentration and a low pH because this situation will favour cadmium accumulation in plants. Running, MANTEQA (a soil chemistry model), revealed that Cd^{2+} and Pb^{2+} concentrations in the soil solution phase of contaminated soils may be reasonably well predicted from the total soil-metal concentration and the soil pH. But the amounts of cadmium taken up by plants or lost through leaching are relatively small compared with the amount added to soils through phosphate fertilizer application. There is therefore a net accumulation of cadmium in soils, particularly in the upper 15 centimeters. The maximum soil Cd concentrations quoted by international guidelines vary between 1 mg/kg and 3 mg/kg. Some fertilizers companies have adopted the limit of 280 mg Cd/kg P representing 25 mg Cd/kg fertilizer in the case of single superphosphate. Frequently pasture soils average 0.44 mg/kg of cadmium, consequently there is potential for high levels of cadmium accumulation even at the 280 mg Cd/kg P limit, relative to the soil content.

Linear modelling of field/pot plant data enables us to draw inferences on Cd accumulation in plants. There is a linear relationship between the total Cd concentration in the vegetative plant (dependent variable) and the plant dry weight, the total Cd concentration in the soil solution and the duration of exposure (time from planting to harvest) (independent variables), as well as a linear relationship between Cd concentration in the reproductive organs /grains and Cd concentration in the vegetative plant organs. No linear relationship was found between Cd concentration in bulk soil and plant organs, or between Cd concentration in the soil solution and grains/reproductive organs. Although Cd uptake influences the total concentration of Ca, Mg, K, P, Mn, Fe, Cu in plants, especially in the early stages of their development, the total Cd concentration in the vegetative plant part is not influenced by the concentration of these metals in the respective organs. For non-toxic Cd concentration levels in the plant organs, the plant mineral status does not influence Cd uptake.

In order to predict cadmium and lead concentrations in air, there are well-established published models that can be used. These models take into account the complex phenomena in the atmosphere and they can make estimations of the deposition rates for Cd and Pd from data on source type, the intensity of atmospheric turbulence, the emission rate, the wind speed, the degree of phenomenon complexity, the surface structure and vegetation properties of the reception area.

Several modeling techniques were used for the submodel construction, such as: meta-regression techniques to compile results from many trials conducted on the absorption of Cd from soil by plants, the uptake of Cd into sheep kidney and liver, the relationship between Cd and Pb intake by humans and standard measures of human toxicity from these two metals as well as linear modeling of endogenous data.

After linking the submodels of the METAL project and conducting sensitivity analyses of the consolidated METAL MODEL we estimated the following influence of factors on Cd concentration in sheep liver and kidney:

Cd concentration in the soil>Number of feeding days>sheep weight >soil
pH =metabolizability of the diet.

The sensitivity analysis also revealed that the rate of cadmium accumulation in kidney is higher than in liver. It was also found that soil pH has no influence on the total Cd concentration in sheep liver although generally it does influence the Cd concentration in the soil solution and thus the total Cd taken up by plants. Cd concentration in the ingested soil by sheep had little influence on Cd accumulation in sheep kidney and liver.

At high levels, cadmium and lead may be toxic to animals and humans, and the toxic effect is chronic rather than acute for Cd. The critical limits of lead accumulation in humans are accurate, but should be modified to take into account the different accumulation potential of men and women. The critical limit of cadmium intake (PTWI) should be increased from 7 to 16 µg/kg of body weight. Human cadmium intake could be effectively reduced by routine removal of the ovine liver and kidney from the food chain in polluted regions.

MAIN PUBLICATIONS FROM THE PROJECT

Györi, Z., Szabó, P., Kovács, B., Prokisch, J. and Phillips, C.J.C. 2004. Cadmium and lead in Hungarian porcine products and tissues. *Journal of the Science of Food and Agriculture* 85, 2049-2054.

Omarova, A. and Phillips, C.J.C. 2006. A meta-analysis of literature data relating to the relationships between cadmium intake and toxicity indicators in humans. *Environmental Research* (in press).

Phillips, C.J.C., Chiy, P.C., and Zachou, E. 2005. The effects of cadmium in herbage on the apparent absorption of elements by sheep, in comparison with inorganic cadmium added to their diet. *Environmental Research* 99, 224-234.

Phillips, C.J.C., Györi, Z. and Kovács, B. 2003. The effect of adding cadmium and lead alone or in combination to the diet of pigs on their growth, carcase composition and reproduction. Journal of the Science of Food and Agriculture. 83, 1357-1365.

Prankel, S. H., Nixon, R. M. and Phillips, C. J. C. 2003. Meta-analysis of experiments investigating cadmium accumulation in the liver and kidney of sheep. *Environmental Research* 94, 171-183.

Prankel, S. H., Nixon, R. M. and Phillips, C. J. C. Implications for the human food chain of models of cadmium accumulation in sheep. *Environmental Research* 97, 348-358.

Tudoreanu, L. and Phillips, C. J. C. 2004. Empirical models of cadmium accumulation in maize, rye grass and soya bean plants. Journal of the Science of Food and Agriculture 84, 845-852.

Tudoreanu, L. and Phillips, C. J. C. 2004. Modelling cadmium accumulation and uptake in plants. Advances in Agronomy 84, 121-157.

Tudoreanu, L., Prankel, S. H., and Phillips, C. J. C. 2006. A Model of Cadmium Accumulation in the Liver and Kidney of Sheep Derived from Soil and Dietary Characteristics (submitted)

Wilkinson, J. M., Hill, J. and Phillips, C. J. C. 2003. The accumulation of potentially toxic metals by grazing ruminants. *Proceedings of the Nutrition Society* 62, 267-277.

In: Environmental Research Advances
Editor: Peter A. Clarkson, pp. 197-207

ISBN: 978-1-60021-762-3
© 2007 Nova Science Publishers, Inc.

Chapter 10

MARINE DEAD ZONES: UNDERSTANDING THE PROBLEM[*]

Eugene H. Buck

ABSTRACT

An adequate level of dissolved oxygen is necessary to support most forms of aquatic life. Very low levels of dissolved oxygen (hypoxia) in bottom-water *dead zones* are natural phenomena, but can be intensified by certain human activities. Hypoxic areas are more widespread during the summer, when algal blooms stimulated by spring runoff decompose to diminish oxygen. Such hypoxic areas may drive out or kill animal life, and usually dissipate by winter.

The largest hypoxic area affecting the United States is in the northern Gulf of Mexico near the mouth of the Mississippi River, but there are others as well. Most U.S. coastal estuaries and many developed nearshore areas suffer from varying degrees of hypoxia, causing various environmental damages. Research has been conducted to better identify the human activities that affect the intensity and duration of, as well as the area affected by, hypoxic events, and to begin formulating control strategies.

Near the end of the 105th Congress, the Harmful Algal Bloom and Hypoxia Research and Control Act of 1998 was signed into law as Title VI of P.L. 105-383. Provisions of this act authorize appropriations through NOAA for research, monitoring, education, and management activities to prevent, reduce, and control hypoxia. Under this legislation, an integrated Gulf of Mexico hypoxia assessment was completed in the late 1990s. In 2004, Title I of P.L. 108-456, the Harmful Algal Bloom and Hypoxia Amendments Act of 2004, expanded this authority and reauthorized appropriations through FY2008.

As knowledge and understanding have increased concerning the possible impacts of hypoxia, congressional interest in monitoring and addressing the problem has grown. The issue of hypoxia is seen as a search for (1) increased scientific knowledge and understanding of the phenomenon, as well as (2) cost-effective actions that might diminish the size of hypoxic areas by changing practices that promote their growth and development. This report presents an overview of the causes of hypoxia, the U.S. areas of most concern, federal legislation, and relevant federal research programs. This report will be updated as circumstances warrant.

---

[*] Excerpted from CRS Report 98-869, dated September 20, 2006.

INTRODUCTION AND BACKGROUND

Hypoxia refers to a depressed concentration of dissolved oxygen in water. While definitions vary somewhat by region, it is generally agreed that hypoxia in a marine environment occurs seasonally when dissolved oxygen levels fall below 2-3 milligrams per liter. Normal dissolved oxygen concentrations in nearshore marine waters range between 5 and 8 milligrams per liter, and many fish species begin having respiratory difficulties at concentrations below 5 milligrams per liter. In extremely low oxygen environments, less tolerant marine animals cannot survive and either leave the area or die [1]. Mortality is especially likely for sedentary species. In addition, spawning areas and other essential habitat can be destroyed by the lack of oxygen. If these conditions persist, a so-called "dead zone" may develop in which little marine life exists [2]. The recovery of marine ecosystems following a hypoxic event has not been extensively studied.

Decreased concentrations of dissolved oxygen result in part from natural eutrophication when nutrients (e.g., nitrogen and phosphorus) and sunlight stimulate algal growth (e.g., algae, seaweed, and phytoplankton), increasing the amount of organic matter in an aquatic ecosystem. As organisms die and sink to the bottom, they are consumed (decomposed) by oxygen-dependent bacteria, depleting the water of oxygen. When this eutrophication is extensive and persistent, bottom waters may become hypoxic, or even anoxic (no dissolved oxygen), while surface waters can be completely normal and full of life. This is encouraged by rising bottom-water temperatures in spring that stimulate increased decomposition by microbes, leading to the development of bottom-water hypoxia.

Eutrophication occurs naturally when offshore winds or surface currents cause cold, nutrient-rich, deep marine waters to rise near coasts, resulting in algal blooms and natural hypoxic events. Many of the hypoxic events along the Pacific and Atlantic coasts are caused by this natural upwelling. However, eutrophication can be increased in intensity or frequency by nutrient loading from nonpoint sources (e.g., runoff from lawns and various agricultural activities including fertilizer use and livestock feedlots), point-source discharge from sewage plants, and emissions from vehicles, power plants, and other industrial sources.

Hypoxic zones frequently occur in coastal areas where rivers enter the ocean (e.g., estuaries). Nutrient-rich fresh water is less dense than saltwater and typically flows across the top of the sea water. The fresh surface water effectively caps the more dense, saline bottom waters, retarding mixing, creating a two-layer system, and promoting hypoxia development in the lower, more saline waters. In the northern Gulf of Mexico, the greatest algal growth in surface waters occurs about a month after maximum river discharge, with hypoxic bottom water developing a month later [3]. Hypoxia is more likely to occur in estuaries with high nutrient loading and low flushing (i.e., low freshwater turnover) [4]. Human activities that increase nutrient loading can increase the intensity, spatial extent, and duration of hypoxic events. Storms and tides may mix the hypoxic bottom water and the aerated surface water, dissipating the hypoxia.

Although the extent of effects of hypoxic events on U.S. coastal ecosystems is still uncertain, the phenomenon is of increasing concern in coastal areas. Several federal agencies are involved in analyzing the problem, including the U.S. Geological Survey (USGS), the National Oceanic and Atmospheric Administration (NOAA), and the U.S. Environmental Protection Agency (EPA). Legislation was enacted by the 105th Congress to provide additional authority and funding for research and monitoring to address these concerns. This authority was extended by the 108th Congress.

HYPOXIC AREAS

Hypoxic episodes have been recorded in all parts of the world, notably in partially enclosed seas and basins where vertical mixing is minimal, such as the Gulf of Mexico, Chesapeake Bay, the New York Bight, the Baltic Sea, and the Adriatic Sea. In March 2004, the U.N. Environment Program's *Global Environment Outlook (GEO) Year Book 2003* reported 146 dead zones where marine life could not be supported due to depleted oxygen levels [5]. Hypoxia is becoming more frequent and widespread in these shallow coastal and estuarine areas [6]. In addition, permanently hypoxic water masses (i.e., oxygen minimum zones) occur in the open ocean, affecting large seafloor surface areas along the continental margins of the eastern Pacific, Indian, and western Atlantic Oceans [7].

About 21% to 43% of the area of United States' estuaries have experienced a hypoxic event; more than half of the affected area is in the Mississippi/Atchafalaya River plume [8] In the Mid-Atlantic region, 13 of 22 estuaries have experienced hypoxic/anoxic events [9]. Of these, the Long Island Sound, Chesapeake Bay, Choptank River, and the New York Bight experience the most serious annual episodes. In the South Atlantic region, hypoxic/anoxic episodes are generally brief, [10] with nearly two-thirds of this region's 21 estuaries experiencing some hypoxia/ anoxia [11]. The Gulf of Mexico region experiences the highest rate of hypoxic/anoxic events, with almost 85% of this region's 38 estuaries experiencing episodes of hypoxia (including the Mississippi/Atchafalaya River plume) [12]. The North Atlantic region is not as prone to hypoxic/anoxic events due to the generally low nutrient input (the result of lower population density) and high tidal flushing. However, areas adjacent to high population density (e.g., Cape Cod Bay and Massachusetts Bay) do experience oxygen depletion. In the Pacific region, hypoxia also occurs near population centers (e.g., San Diego Bay, Newport Bay, Alamitos Bay) or in areas of limited circulation, even where water temperatures are cold (e.g., Hood Canal, Whidbey basin/Skagit Bay) [13].

GULF OF MEXICO DEAD ZONE

The hypoxic zone in the northern Gulf of Mexico is the largest observed in the estuarine and coastal regions of the western hemisphere [14]. First recognized in the early 1970s, it is the largest and most hypoxic area in the United States. In the summer of 1993, following massive Mississippi River flooding, the dead zone covered more than 18,000 square kilometers (an area as large as the state of New Jersey, although pockets of oxygenated water may occur within this boundary), extending westward from the mouth of the Mississippi River to the upper Texas coast [15]. The seasonal shape and extent of the dead zone are mostly a function of the Mississippi/Atchafalaya River plume, the combined outflow from these two major rivers, and the biological processes it influences. This hypoxic zone originates each spring, as melting snow and peak runoff flush nutrients into the Mississippi River system and eventually into the Gulf. The hypoxic zone generally occurs from May to

September, but varies from year to year. After reaching a maximum size of 20,000 square kilometers in 1999, the dead zone covered a much smaller 4,400 square kilometers in 2000. Low velocity winds during the summer result in calm seas that maintain the stratified barrier between surface and bottom water layers. Only during weather disturbances, such as frontal passages, tropical storms, and hurricanes, does vertical mixing of these stratified layers occur. Increased winds and frontal storms in autumn vertically mix the water column, dissipating the hypoxia [16]. In the summer of 1998, this dead zone extended from very near shore (about 10-15 feet water depth) to deeper waters than are normally hypoxic (as much as 160 feet deep off the Mississippi River delta) [17].

Nutrient enrichment is the primary cause of eutrophication, of some algal blooms, and of hypoxia, and is believed to be a major factor in areas such as the northern Gulf of Mexico [18]. The Mississippi watershed drains 41% of the land area of the contiguous 48 states, including most of the farmbelt. Studies of the Mississippi and Atchafalaya Rivers indicate that dissolved nitrogen levels have tripled and phosphorus levels have doubled since 1960, fueling algal growth and the resultant dead zone [19].

Research suggests that fertilizer leaching and runoff from upriver agricultural sources may be the main sources of nutrients. For example, USGS states that 56% of the Mississippi River's nutrient loading results from fertilizer runoff, with an additional 25% of the Mississippi River nitrogen coming from animal manure (municipal and solid wastes account for 6%, atmospheric deposition for 4%, and unknown sources for 9%) [20]. Analysis of cores of sediments underlying the hypoxic area reveals historic information on the Mississippi River watershed, indicating that surface water productivity has increased and bottom water oxygen stress has worsened since the early 1900s, with the most dramatic changes occurring since the 1950s — a change strongly correlated with increased use of commercial fertilizers in the watershed [21]. Although hypoxia occurred in the northern Gulf of Mexico prior to heavy use of artificial fertilizers, human activities have exacerbated and prolonged the hypoxic condition. Other studies also show a direct relationship between the river-born nutrients, the high rates of phytoplankton production, and subsequent Gulf of Mexico hypoxia [22]. However, questions remain as to how much of the river's nitrogen might come from natural soil mineralization, what effects floods have on nutrient transport, and how much nitrogen may be contributed by coastal land loss, estimated at 25 square miles per year [23].

Although studies have found that more than 70% of the total nitrogen transported to the Gulf of Mexico by the Mississippi River originates above the confluence of the Ohio and Mississippi Rivers, [24] focusing on nitrogen runoff per unit area identifies other areas where more concentrated nutrient runoff occurs [25]. Although the lower Mississippi basin (which drains parts of Tennessee, Arkansas, Missouri, Mississippi, and Louisiana) is responsible for only 23% of the nitrogen delivered to the Gulf, some scientists believe that nitrogen removal and/or runoff prevention strategies should focus on this area because of its much greater relative nitrogen contribution [26] and likely more economically efficient nitrogen removal.

Researchers estimate that the benefits of nutrient controls in this lower basin could be twice as effective as implementing them in upstream basins [27]. Workshops and conferences have identified strategies for implementing nutrient controls in the lower Mississippi basin [28].

Many farming interests maintain that evidence has not proven that agricultural practices are the primary contributors to the development of the Gulf of Mexico dead zone [29]. Some farmers dispute that they contribute substantially to creating a dead zone that is as much as

1,000 miles away. They argue that their goal is to keep as much as possible of the applied nutrients on their land, since any nutrients that wash away represent wasted money. On the other hand, it is estimated that as much as half of the applied nutrients are lost to surface or ground water and to the air, resulting in approximately $750 million in excess nitrogen (calculated as fertilizer cost) entering the Mississippi River each year [30].

Impacts of the Gulf of Mexico Dead Zone on Fishing. The Gulf of Mexico supports important, easily accessible commercial and recreational fisheries, bringing in almost $2.9 billion annually in retail sales to Louisiana and supporting almost 50,000 jobs [31]. These highly productive fisheries are the direct result of the input of nutrients from the Mississippi River watershed. To date, no studies have investigated the linkage between fishery declines and hypoxic episodes in the Gulf, but some evidence suggests the dead zone may force fish and shrimp further offshore as well as into shallow nearshore areas (producing what is locally called a "jubilee") that may provide less desirable habitat [32]. Hypoxia increases stress on aquatic ecosystems and may decrease biological diversity in areas experiencing repeated and severe hypoxia [33]. Crowding of marine life into restricted habitat also may lead to indirect consequences through altered competition and predation interactions. In addition, hypoxia may delay or impede the offshore migration of older, larger shrimp, preventing shrimp trawlers from selectively targeting larger shrimp for harvest.

While it is unclear what specific effects the dead zone has on Gulf fisheries, the occurrence of this dead zone may force fishing vessels to change their normal fishing patterns, possibly expending more time and fuel to harvest their catch. One study has concluded that any increase in fishing expenses could drive marginal operators out of business. Other potential impacts on Louisiana fisheries include concentration of fishing effort in other areas, resulting in localized overfishing; damage to essential habitat, and possible decreased future production; shellfish mortality, if hypoxic conditions impinge on barrier island beaches and coastal bay waters; localized mortality of finfish and shellfish in shoreline areas; and decreased growth due to reduced food resources in the sediments and water column [34]. In August 1997, the Louisiana Department of Wildlife and Fisheries initiated a three-year study, funded by the National Marine Fisheries Service (NOAA, U.S. Department of Commerce), to determine the dead zone's impact on commercial fisheries.

CHESAPEAKE BAY

Hypoxic conditions have been recognized in Chesapeake Bay for many years [35]. In 2003, Virginia Institute of Marine Science scientists found a 250-square-mile area of hypoxic water in upper Chesapeake Bay at depths below about 20 feet, from north of Annapolis nearly to the mouth of the Potomac River. The low oxygen levels were attributed to large nutrient inputs, likely carried into the bay by runoff from above-average winter snowfall and spring rains. This runoff was able to pick up nutrients that had accumulated in surrounding soils during four consecutive years of dry weather.

OREGON COAST

Although permanently hypoxic water masses (oxygen minimum zones) affect large seafloor surface areas along the continental margin of the eastern Pacific Ocean, no dead zone events had been reported in the nearshore waters off the Oregon coast prior to 2002. By 2006, the Oregon coastal dead zone was significantly larger, thicker, longer lasting, and lower in oxygen concentration than previous years, extending along the ocean floor from Cape Perpetua (Florence) in the south to Cascade Head (Lincoln City) in the north, as close to shore as the 50-foot depth [36].

The appearance and growth of this dead zone is attributed to fundamental, but not well understood, changes in ocean conditions off the Oregon coast [37].

POLICY AND MANAGEMENT EFFORTS

Since a temporary, yet severe, hypoxic event could result in significant mortality or injury to marine mammals, fish, and other aquatic species, many have deemed better understanding and consistent monitoring of hypoxic phenomena necessary. NOAA initiated the Nutrient Enhanced Coastal Ocean Productivity (NECOP) program in 1989 to study the effects of nutrient discharges on U.S. coastal waters. This study found a clear link between nutrient input, enhanced primary production (i.e., algal and plant growth), and hypoxic events in the northern Gulf of Mexico [38].

In response to a January 1995 petition from the Sierra Club Legal Defense Fund (currently known as Earthjustice Legal Defense Fund) on behalf of 18 environmental, social justice, and fishermen's organizations, the Gulf of Mexico Program [39] held a conference in December 1995 to outline the issue and identify potential actions. Following that conference, Robert Perciasepe, Assistant EPA Administrator for Water, convened an interagency group of senior Administration officials (the "principals group") to discuss potential policy actions and related science needs. Subsequently, this "principals group" created a Mississippi River/Gulf of Mexico Watershed Nutrient Task Force. Additionally, the White House Office of Science and Technology Policy's Committee on Environment and Natural Resources (CENR) conducted a Hypoxia Science Assessment at the request of EPA. The CENR assessment was peer-reviewed, made available for public comment, and submitted to the task force to assist in developing policy recommendations and a strategy for addressing hypoxia in the northern Gulf of Mexico.

In response to an integrated scientific assessment of hypoxia in the northern Gulf of Mexico by the multi-agency Watershed Nutrient Task Force, [40] a Plan of Action for addressing hypoxia was released in January 2001 [41]. Estimates based on water-quality measurements and streamflow records indicate that a 40% reduction in total nitrogen flux to the Gulf is necessary to return to average loads comparable to those during 1955-1970. Model simulations suggest that, short of this 40% reduction, nutrient load reductions of about 20%-30% would result in a 15%-50% increase in dissolved oxygen concentrations in bottom waters. Strategies selected focus on encouraging voluntary, practical, and cost-effective actions; using existing programs, including existing state and federal regulatory mechanisms;

and following adaptive management. A reassessment of progress on implementing this action plan was initiated in 2005 [42].

A key consideration is the level and duration of the necessary reduction in excess nutrients from watersheds. Many agricultural lands have been saturated with nutrients for many years, and it may take a long time to "cycle out" excess nitrogen and phosphorus, even if application rates are reduced [43]. While some believe this problem may have no fast solutions and any management regime considered will need to recognize that progress or improvement may not be apparent for years or even decades, others suggest that improved agricultural practices in efficient application of chemical fertilizers and prevention of soil erosion could yield immediate and measurable benefits. Various policy options for modifying agriculture practices continue to be discussed [44].

Because nonpoint sources are major contributors to the problem at the mouth of the Mississippi River system, many believe the Clean Water Act is the appropriate legal framework for addressing future nutrient inputs. Under §319 of the Clean Water Act, Louisiana and other states have initiated nonpoint-source [45] control programs. These programs seek to combine local, state, and federal agency resources to address pollution from nonpoint sources within each state [46]. To effectively address concerns, however, nonpoint-source programs would need to be encouraged, funded, and implemented throughout the Mississippi River watershed. Under §303 of the Clean Water Act, states must identify water-quality-limited segments of their waters that are not meeting standards, and then establish total maximum daily loads (TMDLs) for each listed water and each pollutant (e.g., nutrients) that is not meeting current water quality standards. In addition, agricultural research and educational outreach/assistance to farmers might complement regulatory efforts.

Congress took note of the hypoxia problem in 1997 when the conference report on FY1998 Department of the Interior appropriations (H.Rept. 105-337) directed the USGS to give priority attention to hypoxia in its FY1999 budget. Near the end of the 105th Congress, provisions of the Harmful Algal Bloom and Hypoxia Research and Control Act of 1998 were incorporated into the Coast Guard Authorization Act of 1998. This measure was signed into law as P.L. 105-383 on November 13, 1998; Title VI authorized appropriations through NOAA to conduct research, monitoring, education, and management activities for the prevention, reduction, and control of hypoxia, harmful algal blooms, *Pfiesteria*, and other aquatic toxins. In 2004, Title I of P.L. 108-456, the Harmful Algal Bloom and Hypoxia Amendments Act of 2004, expanded this authority and reauthorized appropriations through FY2008.

APPROPRIATIONS AND FUNDING

National Ocean Service. Hypoxia research is regularly funded through appropriations to the National Ocean Service — part of the National Oceanic and Atmospheric Administration in the Department of Commerce — in their Extramural Research account under *National Centers for Coastal Ocean Science*. For FY2007, NOAA requested $6 million for extramural research for grants related to harmful algal bloom and hypoxia forecasting.

Agriculture. In the last few years, the U.S. Department of Agriculture's Cooperative State Research, Education, and Extension Service has provided a special research grant of around

$220,000 annually to Iowa State University's Leopold Center for Sustainable Agriculture for a project to define and implement new methods and practices in farming that reduce impacts on water quality and the hypoxia problem in the Gulf of Mexico.

Environmental Protection Agency. In the last few years, the Environmental Protection Agency's Environmental Programs and Management account has provided a specific authorization of around $125,000 annually for the Missouri Department of Agriculture's Hypoxia Education and Stewardship Project. This effort seeks to educate Missouri producers about hypoxia and encourage use of practices that will reduce the amount of nitrogen lost through leaching and/or evaporation.

REFERENCES

[1]   Lisa A. Levin, "Oxygen Minimum Zone Benthos: Adaptation and Community Response to Hypoxia," *Oceanography and Marine Biology: An Annual Review*, v. 41 (2003): 1-45.

[2]   Some molluscs and annelid worms are more tolerant of low oxygen conditions and can survive hypoxic episodes that last many weeks.

[3]   D. Justic et al., "Seasonal Coupling Between Riverborne Nutrients, Net Productivity and Hypoxia," *Marine Pollution Bulletin*, v. 26 (1993): 184-189.

[4]   R. Turner and N. Rabalais, "Suspended Particulate and Dissolved Nutrient Loadings to Gulf of Mexico Estuaries," in *Biogeochemistry of Gulf of Mexico Estuaries*, T. Bianchi, J. Pennock, and R. Twilley, eds. (New York, NY: John Wiley and Sons, 1999), pp. 89-107.

[5]   See *Box 4* at [http://www.unep.org/geo/yearbook/yb2003/089.htm].

[6]   R. J. Diaz and R. Rosenberg, "Marine Benthic Hypoxia: A Review of Its Ecological Effects and the Behavioral Responses of Benthic Macrofauna," *Oceanography and Marine Biology: An Annual Review*, v. 33 (1995): 245-303. (Hereafter referred to as "Marine Benthic Hypoxia")

[7]   John J. Helly and Lisa A. Levin, "Global Distribution of Naturally Occurring Marine Hypoxia on Continental Margins," *Deep-Sea Research, Part I*, v. 51 (2004): 1159-1168.

[8]   N. Rabalais, "Oxygen Depletion in Coastal Waters." *NOAA's State of the Coast Report.* (Silver Spring, MD: NOAA: 1998), National Picture, p. 4, [http://state-of-coast.noaa.gov/ bulletins/html/hyp_09/hyp.html]. (Hereafter referred to as "Oxygen Depletion in Coastal Waters")

[9]   S. Bricker, "NOAA's National Estuarine Eutrophication Survey: Selected Results for the Mid-Atlantic, South Atlantic and Gulf of Mexico Regions," *Estuarine Research Federation Newsletter*, v. 23, no. 1 (1997): 20-21. (Hereafter referred to as "NOAA's Estuarine Eutrophication Survey: Selected Results."); NOAA's Estuarine Eutrophication Survey, vol. 1: South Atlantic Region (Silver Spring, MD: National Ocean Service, Office of Ocean Resources Conservation and Assessment, 1996), p. 50.

[10]   "Oxygen Depletion in Coastal Waters," Regional Contrasts, p. 2.

[11]   "NOAA's Estuarine Eutrophication Survey: Selected Results," pp. 20-21.

[12]   *Ibid.*, pp. 20-21.

[13] "Oxygen Depletion in Coastal Waters," National Picture, p. 5.

[14] N. Rabalais et al., "Consequences of the 1993 Mississippi River Flood in the Gulf of Mexico," *Regulated Rivers: Research and Management*, v. 14 (1998): 161-177.

[15] White House Office of Science and Technology Policy, Committee on Environment and Natural Resources, Hypoxia Work Group, *Gulf of Mexico Hypoxia Assessment Plan* (March 1998), p. 3. (Hereafter referred to as "Hypoxia Work Group.")

[16] In August-September 2005, the winds from Hurricanes Katrina and Rita likely promoted a somewhat earlier dissipation of the dead zone off the mouth of the Mississippi River.

[17] N. Rabalais, press release, Louisiana Universities Marine Consortium (July 27, 1998).

[18] Hypoxia Work Group, p. 2.

[19] R. E. Turner and N. Rabalais, "Changes in Mississippi River Quality This Century — Implications for Coastal Food Webs," *BioScience*, v. 41, no. 3 (1991): 140-147; D. Justic et al., "Changes in Nutrient Structure of River-Dominated Coastal Waters: Stoichiometric Nutrient Balance and Its Consequences," *Estuarine, Coastal, and Shelf Science*, v. 40 (1995): 339-356.

[20] R. H. Meade, ed., *Contaminants in the Mississippi River, 1987-92,* Circular 1133 (Denver, CO: U.S. Geological Survey, 1995), 140 pp.

[21] T. A. Nelson et al., "Time-Based Correlation of Biogenic, Lithogenic and Authigenic Sediment Components with Anthropogenic Inputs in the Gulf of Mexico NECOP Study Area," *Estuaries*, v. 17 (Dec. 1994): 873; B. J. Eadie et al., "Records of Nutrient-Enhanced Coastal Productivity in Sediments from the Louisiana Continental Shelf." *Estuaries*, v. 17 (Dec. 1994): 754-765; N. Rabalais et al., "Nutrient Changes in the Mississippi River and System Responses on the Adjacent Continental Shelf," *Estuaries*, v. 19 (1996): 386-407; S. Gupta et al., "Seasonal Oxygen Depletion in Continental-Shelf Waters of Louisiana: Historical Record of Benthic Foraminifers," *Geology*, v. 24 (1996): 227-230; and R. E. Turner and N. Rabalais, "Coastal Eutrophication Near the Mississippi River Delta," *Nature*, v. 368 (1994): 619-621.

[22] F. H. Sklar and R. E. Turner, "Characteristics of Phytoplankton Production Off Barataria Bay in an Area Influenced by the Mississippi River," *Marine Science*, v. 24 (1981): 93-106; S. E. Lohrenz, M. J. Dagg, and T. E. Whitledge, "Enhanced Primary Production at the Plume/Oceanic Interface of the Mississippi River," *Continental Shelf Research*, v. 7 (1990): 639-664; S. E. Lohrenz et al., "Variations in Primary Productivity of Northern Gulf of Mexico Continental Shelf Waters Linked to Nutrient Inputs from the Mississippi River,"*Marine Ecology Progress Series*, v. 155 (1997): 435-454.

[23] D. Malakoff, "Death by Suffocation in the Gulf of Mexico," *Science*, v. 281 (July 10, 1998): 190-192. (Hereafter referred to as "Death by Suffocation.")

[24] R. Alexander, R. Smith, and G. Schwarz, "The Regional Transport of Point and Nonpoint-Source Nitrogen to the Gulf of Mexico," *Proceedings of the First Gulf of Mexico Hypoxia Management Conference*, Kenner, LA (Dec. 5-6, 1995), pp. 127-133 (hereafter referred to as "Regional Transport"); R. H. Meade, ed., *Contaminants in the Mississippi River, 1987-92,* Circular 1133 (Denver, CO: U.S. Geological Survey, 1995), 140 pp.

[25] Of this total, 39% is contributed by the upper and central Mississippi River basins (which include Minnesota, Wisconsin, Iowa, Missouri, and Illinois), 22% by the Ohio River basin, and 11% by the Missouri River basin.

[26] Nitrogen runoff for the lower Mississippi basin is 2,072 kilograms of nitrogen per square kilometer per year compared to 708 kilograms of nitrogen per square kilometer per year for the upper Mississippi basin and 437 kilograms of nitrogen per square kilometer per year for the Ohio River basin.

[27] "Regional Transport," p. 131.

[28] For example, see C. L. Cordes and B. A. Vairin, eds., *Workshop on Solutions and Approaches for Alleviating Hypoxia in the Gulf of Mexico*, NWRC Special Report 98-02 (Lafayette, LA: U.S. Geological Survey, 1998), 53 p.

[29] C. David Kelly, "Hypoxia Issue Paints a Murky Picture," *Voice of Agriculture, American Farm Bureau* (Sept. 29, 1997), at [http://www.fb.org/index.php?fuseaction=newsroom.focusfocusandyear=1997andfile=fo0929.html].

[30] "Death by Suffocation," pp. 190-192.

[31] Southwick Associates, *The Economic Benefits of Fisheries, Wildlife and Boating Resources in the State of Louisiana* (Arlington, VA: March 1997), 21 pp.

[32] Roger Zimmerman et al., "Trends in Shrimp Catch in the Hypoxic Area of the Northern Gulf of Mexico," *Proceedings for the First Gulf of Mexico Hypoxia Management Conference*, Kenner, LA (Dec. 5-6, 1995), pp. 64-75.

[33] "Marine Benthic Hypoxia," pp. 285-287.

[34] J. Hanifen et al., "Potential Impacts of Hypoxia on Fisheries: Louisiana's Fishery-Independent Data," *Proceedings for the First Gulf of Mexico Hypoxia Management Conference*, Kenner, LA (Dec. 5-6, 1995), pp. 87-100.

[35] Denise L. Breitburg, "Episodic Hypoxia in Chesapeake Bay: Interacting Effects of Recruitment, Behavior, and Physical Disturbance," *Ecological Monographs*, v. 62, no. 4 (1992): 525-546; James D. Hagy et al., "Hypoxia in Chesapeake Bay, 1950-2001: Long-Term Change in Relation to Nutrient Loading and River Flow," *Estuaries*, v. 27, no. 4 (2004): 634-659.

[36] For more information, see [http://www.piscoweb.org/research/oceanography/hypoxia].

[37] Brian A. Grantham, et al., "Upwelling-Driven Nearshore Hypoxia Signals Ecosystem and Oceanographic Changes in the Northeast Pacific," *Nature*, v. 429 (June 17, 2004): 749-754.

[38] NOAA, Coastal Ocean Program Office, *Nutrient-Enhanced Coastal Ocean Productivity, Proceedings of 1994 Synthesis Workshop* (1995), p. 119; see also *Estuaries*, v. 17, no. 4 (Dec. 1994): 729-911.

[39] The Gulf of Mexico Program is a cooperative federal-state effort beginning after Congress, through P.L. 102-178, designated 1992 as the Year of the Gulf of Mexico. For additional information on this program, see [http://www.epa.gov/gmpo/].

[40] See [http://www.nos.noaa.gov/products/pubs_hypox.html].

[41] See [http://www.epa.gov/msbasin/taskforce/pdf/actionplan.pdf].

[42] See [http://www.epa.gov/msbasin/taskforce/reassess2005.htm].

[43] "Death by Suffocation" (citing Don Goolsby, USGS, Denver, CO), pp. 190-192.

[44] For example, see Suzie Greenhalgh and Amanda Sauer, "Awakening the Dead Zone: An Investment for Agriculture, Water Quality, and Climate Change," *World Resources Institute Issue Brief* (February 2003), 24 p.

[45] Originating from land use activities; sediment, organic and inorganic chemicals, and biological, radiological, and other toxic substances are carried to lakes and streams by surface runoff.

[46] D. Sabin and J. Boydstun, "Louisiana Activities and Programs in Nutrient Control and Management," *Proceedings of the First Gulf of Mexico Hypoxia Management Conference,* Kenner, LA (Dec. 5-6, 1995), pp. 196-198.

In: Environmental Research Advances
Editor: Peter A. Clarkson, pp. 209-215

ISBN: 978-1-60021-762-3
© 2007 Nova Science Publishers, Inc.

Chapter 11

NATIONAL PARK SYSTEM: ESTABLISHING NEW UNITS[*]

Carol Hardy Vincent

ABSTRACT

The National Park System includes 390 diverse units administered by the National Park Service (NPS) of the Department of the Interior. Units generally are added to the National Park System by act of Congress, although the President may proclaim national monuments on land that is federally managed for inclusion in the system. Before enacting a law to add a unit, Congress might first enact a law requiring the NPS to study an area, typically to assess its national significance, suitability and feasibility, and other management options. Important areas also are preserved outside the National Park System through programs managed or supported by the NPS.

OVERVIEW OF THE SYSTEM

The National Park System contains 390 units throughout the nation. They are administered by the National Park Service (NPS) of the Department of the Interior (DOI). As of September 28, 2006, the National Park System encompassed 84.6 million acres of land — 79.1 million acres federally owned and 5.5 million acres of private and other public land (e.g. state land) within NPS unit boundaries. Units range in size from less than one acre to more than 13 million acres. Nearly two-thirds of the total acreage is in Alaska.

In 1872, Congress designated Yellowstone as the world's first national park. Subsequently, the nation slowly developed a system of national parks. While some new areas were administered by DOI, others were managed by different agencies. A 1916 law created the NPS within DOI to protect existing and future parks, monuments, and other areas. It charged NPS with promoting and regulating the use of those areas both to conserve them and to provide for their enjoyment by the public. A 1933 executive order furthered the

---

[*] Excerpted from CRS Report RS20158, dated September 28, 2006.

development of a national system by transferring dozens of sites to NPS from other agencies. The General Authorities Act of 1970 made explicit that all areas managed by NPS were part of a single system, and gave all units of the system equal standing with regard to resource protection. Statutes authorizing particular units sometimes provide additional management direction for those units.

Units of the system generally are managed to preserve resources in their natural or historical conditions for the benefit of future generations. Thus, hunting, mining, and other consumptive resource uses generally are not allowed. However, in the laws creating units, Congress sometimes has specified that some of those uses are allowed.

Today, there are more than 20 different designations for units of the National Park System, reflecting the diversity of the areas. As of September 28, 2006, there were 58 units called national parks, the so-called "crown jewels" of the System. Other commonly used titles include national historic sites (78), national monuments (74), national historical parks (42), national memorials (28), national recreation areas (18), and national preserves (18). Some classifications (such as national park) are unique to NPS, while others (such as national recreation area) also are used by other land management agencies [1].

ADDING UNITS BY PUBLIC LAW AND PRESIDENTIAL PROCLAMATION

National Park System units are created by act of Congress, except that national monuments also may be added by presidential proclamation. The Antiquities Act of 1906 (16 U.S.C. 431 et seq.) authorizes the President to create national monuments, on land that is already federally owned or controlled, that contains historic landmarks, historic and prehistoric structures, or other objects of historic or scientific interest [2]. Presidents have designated about 120 monuments since 1906. Congress has subsequently converted many of them, such as the Grand Canyon, to national parks. Most monuments are managed by NPS, with many newer monuments managed by the Bureau of Land Management. (For more information on national monuments, see CRS Report RS20902, *National Monument Issues*, by Carol Hardy Vincent.)

An act of Congress creating a Park System unit may explain the unit's purpose; set its boundaries; provide specific directions for land acquisition, planning, uses, and operations; and authorize appropriations for acquisition and development. Bills to create units generally are within the jurisdiction of the House Committee on Resources and the Senate Committee on Energy and Natural Resources, with appropriations typically contained in Interior Appropriations Acts. In recent years, Congress sometimes has enacted free-standing legislation to add units to the National Park System. Congress also has authorized units as part of omnibus parks and recreation laws containing dozens of recreation-related measures. Measures sometimes are packaged to facilitate broad evaluation of an issue and to expedite consideration. Legislation creating a new unit may be preceded by legislation to authorize an NPS study of the area, as described below.

Provisions of law, together with NPS policies, govern Congress's consideration of measures to create units of the National Park System. In 1998, Congress amended existing law pertaining to the creating units (P.L. 105-391) to standardize procedures, improve the information about potential additions, prioritize areas, focus on outstanding areas, and ensure

federal agencies, state or local governments, Native American authorities, and the private sector. Consideration may be given to technical or financial assistance; other designations, including wilderness, national trail, or national historic landmark; and cooperative management between NPS and another agency. NPS generally will not recommend adding an area to the Park System if another arrangement already provides, or could provide for, sufficient protection and public use. The study must identify the best alternative(s) for protecting resources and allowing public enjoyment. Each study sent to Congress must be accompanied by a letter from the Secretary that identifies the preferred management option for the area, to minimize uncertainty about NPS's position.

ISSUES

The addition of units to the National Park System sometimes has been controversial. Some discourage adding units, arguing that the System is "mature" or "complete," while others assert that the System should evolve and grow to reflect current events, new information, and reinterpretations. A related issue is how to properly maintain existing and new units given limited fiscal and staffing resources. The Bush Administration generally does not support the creation of new park units and the expansion of existing units to focus funds on maintaining current units. The Administration has supported some expansions on the grounds that they could be accomplished for relatively little cost. Supporters of new units have charged that the older units are the most costly. Also, in a departure from the past, the Administration has not recommended to Congress, as part of its annual budget submissions, that areas be studied for possible inclusion in the Park System. The priority has been to complete studies previously authorized by Congress, although the Administration has testified in support of authorizing some new studies [5].

Differences exist on the relative importance of including areas reflecting our natural, cultural, and social history. The adequacy of standards and procedures for assuring that the most outstanding areas are included in the System also has been debated. Critics contend that the System has been weakened by including inappropriate areas, especially where authoritative information was unavailable, incomplete, or disregarded in favor of political considerations. Others counter that there will always be disagreement over the worth of areas, and that recently added areas have been held to the same high standards as older units. Another issue has been whether particular resources are better protected outside the National Park System, and how to secure the best alternative protection.

ALTERNATIVES TO INCLUSION IN THE NATIONAL PARK SYSTEM

It is generally regarded as difficult to meet the criteria and to secure congressional support and funding for expanding the National Park System. Although there may be hundreds or thousands of related inquiries to Congress and the NPS, usually no more than a handful of new units are created each Congress.

Many areas are preserved outside the National Park System. Some of these are protected with recognition or assistance by the NPS. Certain areas that receive technical or financial aid from the NPS, but are neither federally owned nor directly administered by the NPS, have been classified by the NPS as affiliated areas. Affiliated areas are nationally significant but do not meet the other criteria for inclusion in the Park System. Under NPS policy, they are worthy of special NPS recognition or assistance beyond existing programs, are managed in accordance with standards applicable to park units, and are to receive sustained resource protection as detailed in an agreement between the NPS and the non-federal manager of the area. In the past, the affiliated areas have included properties primarily recognized for cultural or commemorative worth. Affiliated areas have been created by act of Congress and by designation of the Secretary of the Interior.

National heritage areas, established by Congress, contain land and properties that reflect the history of their people. Typically, they consist mainly of private properties and may include natural, scenic, historic, cultural, or recreation resources. Conservation, interpretation, and other activities are handled by partnerships among federal, state, and local governments and nonprofit organizations, and for each area Congress has recognized a "management entity" to coordinate efforts. The NPS supports these efforts through technical and financial assistance, and such support is not intended to be permanent. Supporters of heritage areas have asserted that they reduce pressure to add new, costly, and possibly inappropriate areas to the National Park System, while opponents have feared that they could be used to extend federal control over nonfederal land. Differences also have existed over whether to create a comprehensive heritage program containing priorities and standards for establishing heritage areas. (For more information, see CRS Report RL33462, *Heritage Areas: Background, Proposals, and Current Issues*, by Carol Hardy Vincent and David Whiteman).

Some programs give places honorary recognition. Cultural resources may be listed by the NPS in the National Register of Historic Places, as meriting preservation and special consideration in planning for federal or federally assisted projects. The Secretary of the Interior may designate natural areas as national natural landmarks, and cultural areas as national historic landmarks. National parks, monuments, and other areas of international worth may, at the request of the United States, be recognized by the United Nations as world heritage sites or biosphere reserves. The Congress, or the Secretary of the Interior, may designate rivers as components of the National Wild and Scenic Rivers System, and trails as part of the National Trails System.

The NPS supports local and state governments in protecting resources. The agency may provide grants for projects (including acquisition and development of recreational facilities), and technical assistance (for conserving rivers, trails, natural areas, and cultural resources). In addition to this range of NPS programs, resources are protected by the private sector, state and local governments, and other federal agencies.

REFERENCES

[1]   A brief definition for each classification, together with a description of each unit of the System, is included in U.S. Dept. of the Interior, National Park Service, Office of Public Affairs and Harpers Ferry Center, *The National Parks: Index 2005-2007* (Washington, DC: GPO, 2005).

[2]   Extensions or establishment of monuments in Wyoming require the authorization of Congress (16 U.S.C. 431a), and withdrawals in Alaska exceeding 5,000 acres are subject to congressional approval (16 U.S.C. 3213).

[3]   See 16 U.S.C. 1a-5 for provisions of law.

[4]   U.S. Dept. of the Interior, National Park Service, *Budget Justifications and Performance Information, Fiscal Year 2007*, p. Const-82-83.

[5]   *Ibid.,* p. Const-82-83.

In: Environmental Research Advances
Editor: Peter A. Clarkson, pp. 217-223

ISBN: 978-1-60021-762-3
© 2007 Nova Science Publishers, Inc.

Chapter 12

ENVIRONMENTAL QUALITY INCENTIVES PROGRAM (EQIP): STATUS AND ISSUES*

Carol Canada and Jeffrey Zinn

ABSTRACT

The Environmental Quality Incentives Program (EQIP) provides farmers with financial and technical assistance to plan and implement soil and water conservation practices. EQIP was enacted in 1996 and amended by the Farm Security and Rural Investment Act of 2002 (Section 2301 of P.L. 107-171). It is a mandatory spending program (i.e., not subject to annual appropriations) administered by the Natural Resources Conservation Service. EQIP is guaranteed a total of $10.0 billion from FY2002 through FY2010 from the Commodity Credit Corporation (CCC), making it the largest conservation financial assistance program [1]. Issues about EQIP that Congress may explore as it starts to consider the next farm bill include (1) reducing the pending backlog of applications, (2) measuring the program's accomplishments, and (3) using EQIP to address specific topics or needs in specified locations.

BACKGROUND

EQIP is the principal source of financial assistance (cost-sharing payments and incentive payments) for agricultural producers who wish to implement soil and water conservation practices. It also provides participants with technical assistance. Participation is voluntary. EQIP was created by the Federal Agriculture Improvement and Reform Act of 1996 (P.L. 104-127, April 4, 1996) and replaced four conservation programs repealed in the same law. These were the Great Plains Conservation Program, the Agricultural Conservation Program (ACP), the Water Quality Incentives Program, and the Colorado River Basin Salinity Control Program.

---

* Excerpted from CRS Report RS22040, dated August 15, 2006.

EQIP PROGRAM TODAY

EQIP is administered by the U.S. Department of Agriculture's Natural Resources Conservation Service (NRCS) under a final rule, published in the May 30, 2003, *Federal Register* [2]. This rule implements amendments to the program enacted in the Farm Security and Rural Investment Act of 2002 (Section 2301 of P.L. 107-171, May 13, 2002), commonly referred to as the 2002 farm bill. EQIP's legislative mandate is to optimize environmental benefits. NRCS implemented this by establishing national priorities to reflect the most pressing natural resource needs and emphasize off-site benefits to the environment. NRCS considers these national priorities in allocating funds to states and establishing cost-share and incentive payment levels. The current national priorities are:

- Reduction of nonpoint source pollutants in impaired watersheds (consistent with Total Maximum Daily Loads, [3] or TMDLs), reduction of groundwater contamination, and reduction of pollution from point sources;
- Conservation of ground and surface water resources;
- Reduction of emissions that contribute to air quality impairment violations of National Ambient Air Quality Standards;
- Reduction of soil erosion and sedimentation from unacceptable levels on agricultural land; and
- Promotion of at-risk species habitat conservation.

NRCS also has identified state and local priority natural resource concerns that support the national priorities. These state and local priorities are used as guidelines by states when selecting which producers will receive EQIP assistance.

HOW EQIP WORKS

Producers with eligible land can apply by submitting an EQIP plan that describes the conservation and environmental purposes that the producer will achieve by using one or more USDA-approved conservation practices. Eligible land includes cropland, rangeland, pasture, private non-industrial forest land, and other lands as determined by USDA. Of the total authorized annual spending, 60% is allocated to livestock (both confined and grazing) practices.

USDA-approved conservation practices may involve structures, vegetation, or land management. Structural practices include the establishment, construction, or installation of measures designed for specific sites, such as animal waste management facilities, livestock water developments, and capping abandoned wells. Vegetative practices involve introduction or modification of plantings, such as filter strips or trees. Land management practices require site-specific management techniques and methods, such as nutrient management, irrigation water management, or grazing management.

NRCS partners with units of federal, state, and local governments, and interest groups to coordinate information and resources that address both local and national priorities. In addition, producers can receive technical assistance from NRCS, state, local, or federal conservation offices, or approved third party providers to develop an EQIP plan and, after approval, to implement it. The local conservation district will review the plan and then decide whether or not to select the plan for EQIP funding. If approved, USDA will provide cost-share payments or incentive payments to help the producer offset the cost of the practice. Participants are eligible to receive cost-share payments for both constructing structures and implementing land management practices. In addition, they may be eligible to receive incentive payments from implementing certain higher-priority practices, such as developing comprehensive nutrient management plans.

Contracts are in effect from one to ten years. They are capped at $450,000 each, and total payments to a person or entity over any six-year period is also limited to $450,000. Individuals or entities with an average annual adjusted gross income (AGI) of $2.5 million for the three years prior to the contract period are ineligible unless they received 75% of their AGI from farming, ranching, or forestry. USDA will pay up to 75% of the projected cost of each practice; however, limited resource producers and beginning farmers and ranchers [4] can receive up to 90% of the cost. Initial payments are made the year in which the contract is signed.

EQIP FUNDING

The 2002 farm bill (Section 2701) funded EQIP at a total of $6.1 billion in mandatory funds from the CCC through FY2007. Of the $6.1 billion, $5.8 billion is to be used to fund the cost-share portion of EQIP with the remainder for the new programs discussed in the next section. The program, which had been authorized at $200 million annually before FY2002, was to receive $400 million in FY2002, increasing up to $1.3 billion in FY2007. A main justification for this increase was to respond to the large backlog of producer demand that had been documented during the 2002 farm bill debate. As a mandatory program, it would have automatically received the authorized amounts if Congress had not acted to reduce funding every year through annual appropriations legislation. Funding was authorized at a total of almost $4.4 billion between FY2003 and FY2006, but Congress limited it each year, providing a total of $3.952 billion. This reduction is almost 10% below the authorized total [5].

Table 1. EQIP Allocations for the Three Largest Recipients,
FY1998-FY2006 ($ million)

State	1998	1999	2000	2001	2002	2003	2004	2005	2006
National Total	198.2	174.0	176.6	199.9	387.0	626.7	908.3	991.9	1,013.3
Texas	16.3	13.4	13.3	15.2	28.7	57.7	78.6	90.0	91.3
California	7.8	8.1	7.8	9.2	19.1	48.6	57.0	62.1	62.9
Colorado	6.4	7.5	7.0	7.1	14.4	25.6	36.9	39.2	41.2

Source: USDA, NRCS.

EQIP funding levels were amended by Section 1203 of the Deficit Reduction Act of 2005 (P.L. 109-171), which limited funding to $1.27 billion in FY2007. This legislation also extended the authorization through FY2010, providing $1.27 billion in FY2008 and FY2009, and $1.3 billion in FY2010.

As shown in Table 1, the same three states have received the most EQIP funds each year. The leading state is Texas, which has received approximately 9% of the total in each of the past four years. The next highest states have been California followed by Colorado each year. In FY2006, the next four states receiving the largest amounts were Minnesota ($32.0 million), Nebraska ($31.8 million), Montana ($31.7 million), and Kansas ($30.8 million).

NEW PROGRAMS UNDER EQIP

Three sub-programs, enacted in the 2002 farm bill, are implemented through EQIP. One of these, the Competitive Conservation Innovation Grants (CIG), is intended to leverage federal investment, stimulate innovative approaches to conservation, and accelerate technology transfer in environmental protection and agricultural production. CIG is authorized from FY2003 through FY2006 at an unspecified annual funding level. Grants must not exceed 50% of the project cost, with non-federal matching funds provided by the grantee. NRCS established national, state, and Chesapeake Bay watershed components, and allocated up to $10 million, up to $5 million, and up to $5 million, respectively, in FY2006. For funding in earlier years, CIG was not implemented in FY2003, then awarded $14.3 million in FY2004 and $22.0 million in FY2005.

A second sub-program, the Ground and Surface Water Conservation (GSWC) program, provides cost-share and incentive payments to producers where the assistance will result in a net savings in ground or surface water resources in the producer's agricultural operation. Funding is authorized as a separate amount in addition to EQIP at $25 million in FY2002, $45 million in FY2003, and $60 million annually from FY2004 through FY2007. Congress limited funding to $51 million in FY2004, FY2005, and FY2006 [6]. In the third sub-program, producers in the Klamath Basin in California and Oregon continue to receive money from a separate and additional $50 million authorization, to be provided "as soon as practicable" to install conservation practices and manage irrigation waters [7].

SELECTED POLICY ISSUES

EQIP enjoys widespread support in the farm community and in Congress as it continues to be the major source of financial assistance to help producers implement conservation practices that address specific resource and environmental problems. Major issues that might be discussed in anticipation of the next farm bill include the backlog of interest that is not being met at current funding levels, assessing more precisely what is being accomplished through the EQIP program, and using EQIP to address specific topics or needs in specified locations.

APPLICATIONS BACKLOG

The number of applications for EQIP funding each year has been large enough that NRCS has not been able to clear the backlog, even with much higher funding levels. As show in Table 2, the gap between the supply of funds and the demand for them expanded rapidly in FY2002 and FY2003 after remaining fairly constant earlier.

Table 2. EQIP Applications and Contracts

Fiscal Year	Total Applications	Contracts (% of applications)	Backlog Applications
1998	54,816	19,758 (36.0%)	35,058
1999	51,877	18,847 (36.3%)	33,030
2000	53,961	16,249 (30.1%)	37,712
2001	47,461	17,684 (37.3%)	29,777
2002	90,312	19,817 (21.9%)	70,495
2003	204,313	30,251 (14.8%)	174,062
2004	181,807	46,413 (25.5%)	135,394
2005	82,114	49,406 (60.2%)	32,708

Source: NRCS, USDA.

In FY2004 and FY2005, the number of contracts continued to grow, and the backlog decreased to levels not experienced since FY2001. Between FY1998 and FY2001, NRCS awarded contracts to approximately one-third of the applicants each year, but that declined to 22% in FY2002 and to only 15% in FY2003, then increased to 25% in FY2004. NRCS awarded contracts to more than half of the applicants in FY2005. The backlog is still large even with the additional funding, and if the backlog was a major justification for higher funding in the 2002 farm bill debate, it could be a strong argument in the next farm bill debate. However, from FY2004 to FY2005, a very large number of backlog applications were not updated (for unknown reasons), and are therefore no longer considered. As a result, the total backlog declined by almost 103,000, even though only 50,000 applications were funded.

While the backlog is large, detailed information is not currently available on the characteristics of those applications. Of particular interest may be whether there are any agricultural regions where a much smaller portion of applications are being funded, whether some practices are more likely to remain in the backlog than others, and whether some applications stay in the backlog for a much longer time period than others.

MEASURING EQIP ACCOMPLISHMENTS

NRCS can provide considerable information about EQIP contracts, including which conservation practices are being installed, and their design and maintenance standards. However, relatively little is known about what is actually being accomplished through EQIP contracts, or how enduring those accomplishments are after the contract ends. Among the questions that NRCS is trying to address for all its conservation activities, including EQIP, are how to evaluate performance, how to measure environmental changes, how to evaluate cost effectiveness, which methods to use to identify environmental effects, and which types of data should be collected to measure output. NRCS has recently initiated a national review of its conservation accomplishments called the Conservation Effects Assessment Project (CEAP) to develop better answers to all these questions. While it has committed several million dollars annually to this effort, few results will be available before the next farm bill debate begins.

Regarding EQIP specifically, NRCS has proposed to periodically review state-prepared reports to determine how the program is being delivered at the state and local level. NRCS will require states to prepare reports describing EQIP implementation and accomplishments tied to performance measures. This information will be available to the public from the NRCS website. Of particular interest may be livestock production practices, which receive 60% of total funding each year. Policy makers may seek more information about the development and adoption of Comprehensive Nutrient Management Plans, especially by Confined Animal Feeding Operations, referred to as CAFOs. CAFOs are large livestock operations; the minimum number of animals varies with the type of animal. Some have expressed concern that the effects of CAFOs on the environment and public health have not been adequately assessed, and may seek to address those concerns in the next farm bill.

TARGETING EQIP

A small portion of EQIP funding is now targeted, as a result of provisions enacted in 2002, to the three sub-programs described above. One, the Klamath Basin, is targeted to a specific area, and the other two are targeted to specified topics, ground and surface water conservation, and innovative conservation technologies. A question that may be addressed in the farm bill debate is whether more sub-programs should be created, and if so, what topics will these programs address and how much money will be committed to them. Interests may promote many different topics for such programs, but if funds for these sub-programs are taken out of the general EQIP program, that may attract opposition, especially if EQIP funding remains at currently authorized levels, as is widely anticipated as a preferred outcome in the current budget environment.

REFERENCES

[1]   The CCC is administered by a Board of Directors from agencies of the Department of Agriculture. It has no staff, and all work done on its behalf is performed by staff of agencies within USDA. For EQIP, NRCS provides the staff.

[2]   "Environmental Quality Incentives Program Final Rule," *Federal Register*, vol. 68, no. 104, May 30, 2003, pp. 32337-32355.

[3]   For more information on TMDLs, see CRS Report 97-831, *Clean Water Act and Total Maximum Daily Loads of Pollutants*, by Claudia Copeland.

[4]   A limited resource producer or rancher has direct or indirect gross farm sales of not more than $100,000 in each of the previous two years (adjusted for inflation) and a total household income at or below the national poverty level for a family of four, or less

than 50% of county median household income (determined using Commerce Department data), in the previous two years. A beginning farmer or rancher is an individual or entity who has farmed for less than 10 years.

[5]   The percentage reduction was very similar before the 2002 farm bill, as the 1996 farm bill authorized funding at a total of $1.0 billion between FY1997 and FY2002, but appropriators limited it to a total of $897 million, reducing it by slightly more than 10% of the authorized total.

[6]   In FY2005, GSWC provided approximately $63 million in funding to producers (includes $51million congressionally limited funding and $12 million through another farm bill provision that requires a minimum total funding to each state in support of regional equity).

[7]   In addition, the Administration has undertaken other initiatives, including pilot projects that provide market-based incentives for water quality and target small and limited resource farmers.

INDEX

B

C

G

N

T

X

Y

Z